Y0-ACF-868

THE SYSTEMS ANALYSIS WORKBOOK

A Complete Guide to Project Implementation and Control

Second Edition

Also by Robert D. Carlsen:

The Encyclopedia of Business Charts
with Donald L. Vest

Handbook of Personnel Administration Forms and Formats
with James McHugh

Handbook of Sales and Marketing Forms and Formats
with James McHugh

Handbook of Production Management Forms and Formats
with James McHugh

Handbook of Research and Development Forms and Formats
with James McHugh

Handbook of Construction Operations Forms and Formats
with James McHugh

THE SYSTEMS ANALYSIS WORKBOOK

A Complete Guide to Project Implementation and Control

Second Edition

Robert D. Carlsen
and
James A. Lewis

Graphics by
Donald L. Vest

Prentice-Hall, Inc.
Englewood Cliffs, N.J.

Prentice-Hall International, Inc., *London*
Prentice-Hall of Australia, Pty. Ltd., *Sydney*
Prentice-Hall of Canada, Ltd., *Toronto*
Prentice-Hall of India Private Ltd., *New Delhi*
Prentice-Hall of Japan, Inc., *Tokyo*
Prentice-Hall of Southeast Asia Pte, Ltd., *Singapore*
Whitehall Books, Ltd., *Wellington, New Zealand*

Ninth Printing . . . . . June, 1980

Library of Congress Cataloging in Publication Data

Carlsen, Robert D.,
The systems analysis workbook.

Includes index.
1. System analysis. I. Lewis, James A. joint author. II. Title.
T57.6.C36 1979 658.4'032 78-15021
ISBN 0-13-881565-8

Printed in the United States of America

About the Authors

Robert D. Carlsen operates a nationwide consulting and systems service company headquartered in Southern California, and specializes in the design and development of management, administrative and information retrieval systems. Mr. Carlsen's extensive career includes responsibilities as business executive and manager, management consultant, systems analyst and methods engineer. He also conducts a series of popular seminars on systems analysis.

Significant systems projects in which the author has applied the *Workbook* approach include a systems analysis of the stock clearing system for the American Stock Exchange, development of a new over-the-counter trading system for Merrill Lynch, Pierce, Fenner and Smith, and a new base-wide management information system for the U.S. Naval Submarine Base, New London, Connecticut. As Director of Operations, he developed all the basic planning for a joint Rockwell International/Avondale Shipyards ship production project for the U.S. Navy. The author has also established and managed the Management Information Center at Litton Industries Guidance and Control Systems Division. The author has been President of Computer Real Time Systems, Inc., and directed development of advanced modular information management systems.

James A. Lewis's systems background includes analysis and design, software marketing and consulting. He has participated in a number of management information systems projects during various stages of their planning and development, including projects involving the McDonnell Douglas Corporation, Grolier Incorporated, the Jet Propulsion Laboratory and Falcon Plastics Division of Bioquest. He worked directly with Mr. Carlsen in the application of the *Workbook* approach in the design of the Project Grant Information System for the U.S. Office of Education and other projects. As a senior systems development analyst at Rockwell International, he worked on numerous other Office of Education projects, including the Educational Resources Information Center (ERIC) program. He has worked for the Southern California Edison Company in improving systems of operation, and more recently has been News Editor for *Wassaja*, a national newspaper published by the American Indian Historical Society.

Mr. Lewis is a graduate of the California Institute of Technology and did graduate work at the Institute for Communications Research at Stanford University. He is a member of the American Society for Information Science and the Society for Technical Communications. His background also includes management of logistics and technical publications in the electronics industry, and instruction in technical writing and editing in the Los Angeles public schools.

What This *Second Edition* of the Workbook Will Do for You

"The authors have assembled what could become the classic textbook for systems engagements. . . ." That's what was reported in the CPA Journal of December 1973 regarding the first edition.

Indeed, the first edition of the *Systems Analysis Workbook* was highly praised as a practical, complete guide to analyzing existing systems of operation, and designing new, improved systems. It was described by reviewers as ideally suited for either manual or computer-related systems development projects. The authors were represented as having "given the subject the 'full treatment' in this well-prepared, well-organized and well-documented book."

Now comes the new, expanded *Second Edition*. It contains all the valuable material that was in the first edition, plus entirely new sections and chapters:

- Explaining systems development projects to management . . .
- Designing decentralized system applications which utilize minicomputers . . .
- Documenting the system with comprehensive operation instructions for its use, and detailed descriptions and specifications for trouble-shooting, modification and improvement purposes . . .
- Developing input and output forms, and business forms of all types, that greatly reduce the probability of errors in system operation . . .
- Monitoring and evaluating a system's operation, once it is implemented, to spot problems and take corrective actions.

As with the original edition, the *Workbook* provides a systematic method for analyzing both simple and complex projects. It provides work sheets, step-by-step guides and instructions and master checklists for:

- Collecting data.
- Determining real information and reporting requirements.

- Describing the existing system.
- Developing improvements and designing new systems.
- Evaluating the recommended changes, including cost/benefit factors.
- Planning the new or improved system's programming (if it's a data processing system) and installation (implementation).

Decentralized data processing systems which utilize sophisticated minicomputers are taking over many of the functions that previously could not be automated or were done in large centralized computer centers. This trend has had a large impact on planning, designing and implementing new systems. New questions must be answered as to equipment selection and other matters. The *Second Edition* contains new material that covers this subject.

The forms used in a system have an often-overlooked impact on system efficiency. Forms are the primary means of transferring information within the system, of communications between operating groups, between people and the computers, and from the computer to the users. A high error rate can be attributed to inefficient forms. The *Workbook* now has checklists and work sheets for evaluating the vital subject of communicating information within the system.

How much work can the system handle? What are a system's limitations? How can critical bottlenecks be identified and measured? How should a system be documented? What is involved in monitoring and evaluating an operating system? The new edition of the *Workbook* addresses these subjects with new sections, chapters and new work sheets and checklists.

The basic *Workbook* concept has not changed. The *Workbook* is still a flexible tool that can be as big or as small as needed, depending on the scope of the project. It still allows teams of analysts of different backgrounds to work together and generate a uniform and integrated systems package.

What has been done is an updating of the *Workbook* for *today's* systems.

Robert D. Carlsen
James A. Lewis

Acknowledgments

Any number of people have provided the authors with technical assistance and much needed encouragement, and to them our sincere gratitude. In particular we would like to acknowledge and thank Mr. H. Snowden Marshall for the hours of editorial and proofreading effort spent in reviewing the original text during its production, and Diane Carlsen for her diligent and invaluable work at the typewriter. For editorial work and typing of the *Second Edition*, we must thank Irene Nakamura.

Contents

1
The Workbook—An Efficient Approach to Systems Analysis

A system is an operation or combination of operations performed by people and, possibly, machines to carry out a specific business activity. This might be a total system that considers all the factors in the entire operation of an enterprise, or it might be a subsystem of that total.

A systems analysis is a study of one of these systems or subsystems. The purpose is to evaluate the system in terms of one or more of the following factors . . . efficiency, accuracy, timeliness, economy, and productivity . . . and to design a new or improved system. The design should eliminate or minimize deficiencies and improve the overall operations. Basically the systems analyst who performs the study is concerned with three things. First, he must consider what is currently being done. Second, he must develop a method for what should be done. Finally, he must plan for the new design's application and for implementation of the system. Systems analysis is the first step in the development of a successful automated computer system, but the results of a systems analysis do not necessarily have to result in an automated system.

SYSTEMS ANALYSIS AND THE WORKBOOK

This book presents a uniform, systematic technique for aiding in the performance of all steps of a systems analysis through use of a standardized Workbook. The Workbook contains a set of work sheets and checklists flexible enough in design and selection that they can be added to, deleted, or modified to meet the special requirements of most systems analysis projects, large or small.

The use of a workbook is, of course, but one of several acceptable techniques for conducting a systems analysis. The method used on any given project should be the one best suited to the requirements of that project. The experience of the analyst or analysts assigned, and the scope of the project, are important factors in selecting that approach. An experienced analyst, working on a system improvement task of limited scope, may be able to go directly to the preparation of a flowchart, for instance. If the system being studied is a manual one, a procedure may be written directly from the flowchart and, when this procedure is approved and instituted, "analysis, design, and implementation" have all been accomplished with a minimum of superfluous effort.

Regardless of the approach taken, a review of the Workbook described in this volume will be of value in understanding systems analysis and will serve as a guide in planning any systems project. To that end it has been made as inclusive as possible, even though it is realized that seldom will all of the work sheets be required except, possibly, on large, extremely comprehensive systems analysis projects. Moreover, the work sheets that are used may have to be modified to suit the needs of a particular project.

For many projects the workbook approach enables the analyst to do a better job under a variety of circumstances. For one thing, it makes the analyst more productive, assuring him that in his study effort, important factors are not overlooked. It provides—for a team of analysts—a uniform way of working together. It serves as a continuous, ever-evolving, and always-up-to-date progress report for management. It trains and guides the first-time analyst by showing what needs to be looked at, what types of data need be collected, and the form best suited for communicating ideas for improvements.

Systems analysts working in the field today, whether as employees of a company and concerned with departmental improvements, or working independently as consultants, have learned their craft in a variety of manners. The training ground for some has been as programmers, coding instructions for data processing equipment. It is reasonable to expect systems analysts with this type of background to be especially skilled in designing systems that are readily producible (easily programmable). Other systems analysts have arrived at their position by working first as methods engineers involved in all types of system improvement studies, then drifting more and more toward just those types of improvements involving automatic data processing. It is reasonable to expect analysts with this type of background to be especially skillful in considering the human elements of the total system. A third category of systems analysts concerns those having unique special knowledge and skills such as in law, or in library work, or in mathematics, or in any other of a thousand specialties. It is reasonable to expect analysts with specialty backgrounds to be especially skilled in developing useful systems involving their own specialties.

On any given project one is likely to find systems analysts from a wide variety of backgrounds. In situations like this, the Workbook is especially useful in helping to take advantage of the unique skills that people from differing backgrounds can bring to a project. In order to best understand this and other important aspects of the usefulness of the Workbook, it is best to first examine the environment in which it is used.

TYPES OF SYSTEMS

In the broadest sense, the entire universe is a system and everything else is a subsystem. Each subsystem, whether biological, environmental, man-produced, or other, consists of an input-process-output-feedback cycle which is interconnected with other subsystems in a complex pattern.

A narrow section of that vast mosaic of subsystems is covered in this book. Here the concern is primarily with data processing subsystems that are manual or utilize a combination of manual and electrical or electronic processing. Specifically, these consist of:

Manual Data Processing. Uses pencils, paper, typewriters, slide rules, adding machines and other such devices.

Electrical Accounting Machine (EAM) Processing. Uses manual tasks plus sorters, collators, calculators and other punched-card processing devices (which are often controlled by wiring boards).

Electronic Data Processing (EDP). Uses manual tasks plus program-controlled large computers, medium-sized computers or minicomputers which operate a variety of input, output and data storage devices and peripheral equipment.

ELEMENTS OF A SYSTEM IMPROVEMENT PROJECT

The diagram shown in Figure 1 illustrates the basic functions involved in the development of a new or improved system of operation. The systems analyst is first concerned with the analysis of the existing system, then the design of a new or improved system. At this point the system is procedurized, in detail. For those portions of the system that are manual (the human elements) this means in many cases, and quite literally, the writing of procedures. For the data processing portion of the system the

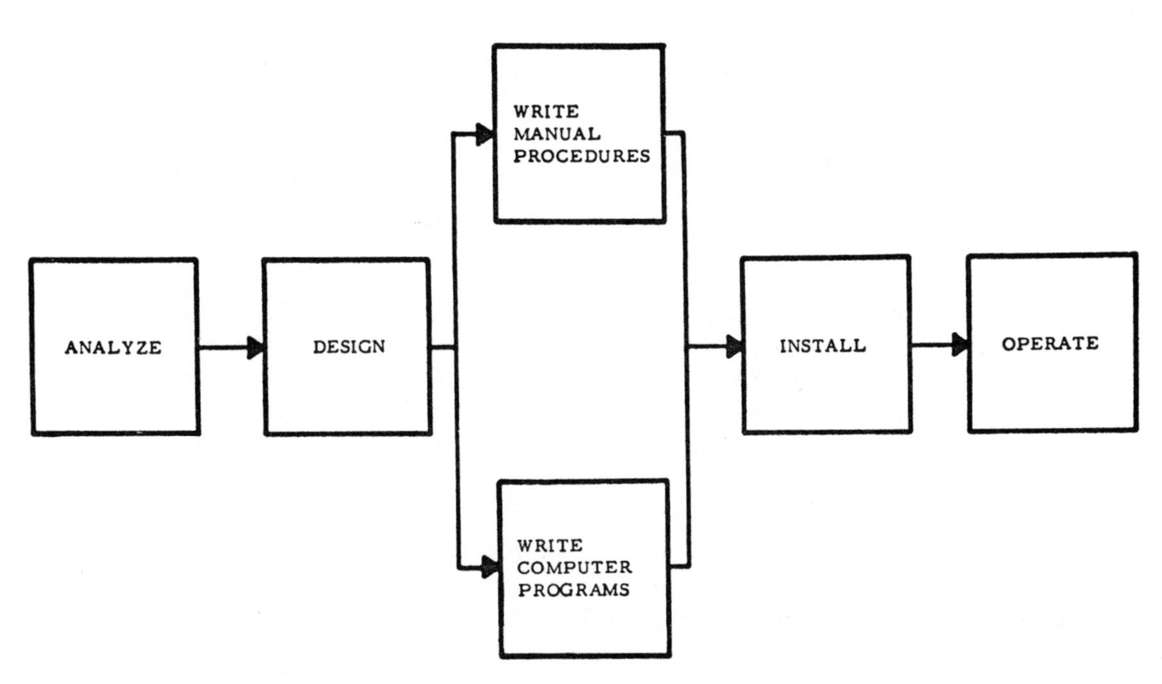

FIGURE 1: Elements of a System Improvement Project

identical type action takes place with the exception that these procedures (called "programs") are instructions for data processing equipment rather than for humans. When the procedures and programs are completed the system is installed and, once tested and proven, becomes operational.

PHASES OF SYSTEMS ANALYSIS

A systems analysis project is usually thought of as occurring in two separate phases, as shown in Figure 2. The first phase involves both the study of the existing system and the design of the new or improved system. The second phase involves implementing the new or improved system. This means writing the detailed procedures and data processing programs, conducting various types of tests, and installing the new system.

There is a very practical reason for having a project divided into these two phases. This point is also illustrated in Figure 2, where it shows the client, or department manager, in a position to exercise approval before the start of each of the two phases. In the first instance, he exercises approval of the study objectives, and approval of the cost of the analysis and design effort. After the new or improved system design is complete, he again is in the position of exercising approval.

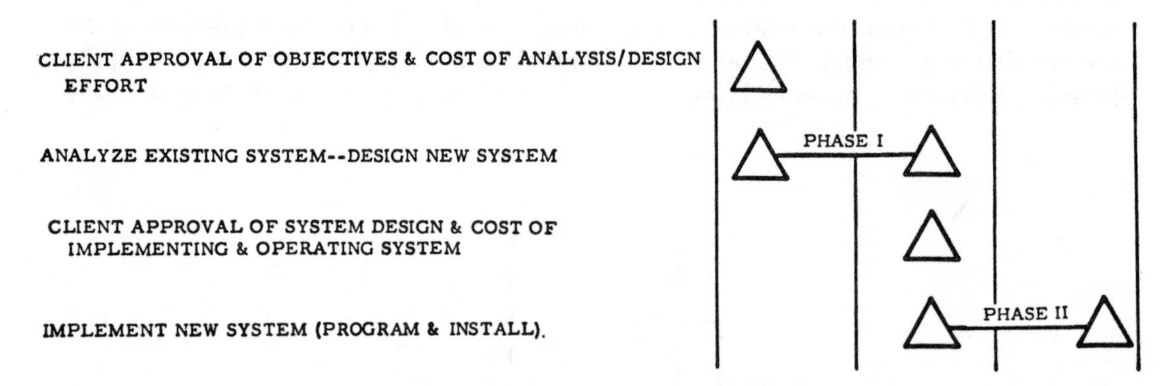

FIGURE 2: Phases of Systems Analysis

In the second instance the management-decision information available to him includes a final system design, a detailed breakdown of the costs required to implement the new system, and an accurate estimate of the cost to operate it. Prior to any further expenditure of time or money, he is in the position to veto the project, alter basic precepts or the direction of the project, or approve it, as is. Furthermore, at this point he is also able to evaluate the analysts' ability to plan and control their own work, based on their performance in the first phase of activity.

What this two-phase approach offers, then, is that the client, or department manager, is not put in the position of buying a "pig in a poke." In truth, there is no systems analyst who can accurately estimate the implementation and operational costs of a system that has not yet been designed. An analogy that seems illustrative would involve the efforts of someone trying to estimate the cost, schedule, and benefit factors for a new home to be constructed, without the benefit of the architect's design or blueprint.

STEPS IN SYSTEMS ANALYSIS

The separate steps involved in a large-scale systems analysis project (Phases I and II, included) are illustrated in Figure 3. It must be emphasized that not all systems analysis situations require such depth and scope. By reviewing this type of detail, however, the analyst should be able to better understand the form and pattern of an analysis, with the understanding that it is always necessary to scale his efforts according to the project he is confronted with.

The detailed steps of an analysis have been presented on the chart in such a manner as to emphasize the two concurrent activities that must take place: System Development, and Project Planning and Control. The numbers to the left of each specified step indicate the sequence in which these steps will be described in the following paragraphs.

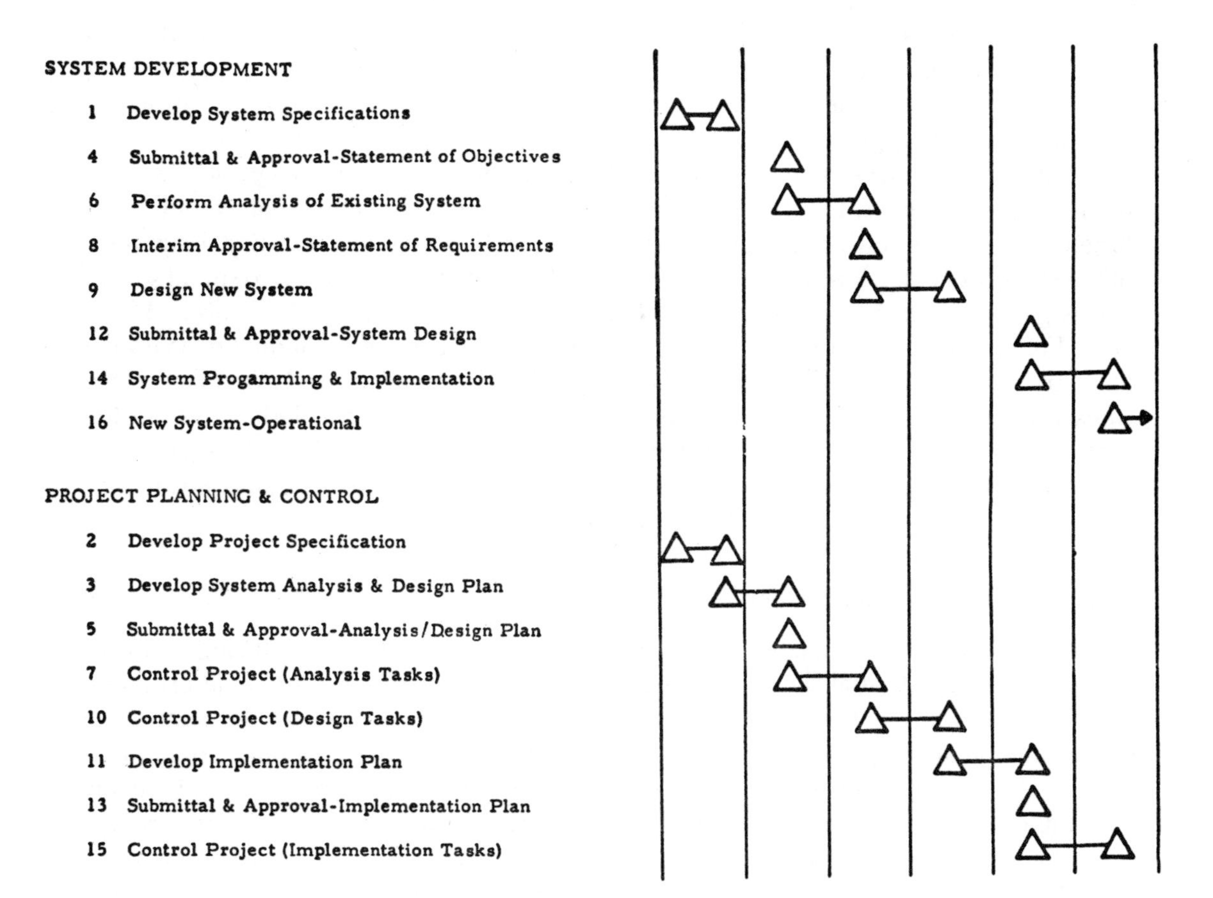

FIGURE 3: Steps in Systems Analysis

1 Develop System Specification

The first step in any systems analysis project involves communicating with the potential customer (this could be an in-house department), for the purpose of defining the purpose of the analysis. This involves determining problems that exist in the current system, specifying system objectives (goals), and listing possible system constraints or limitations. This latter category might be in the form of client-imposed limitations regarding the type or brand of data processing equipment that might be utilized in a new or improved system.

At this point the analyst also concerns himself with what is commonly called "Pre-Analysis Familiarization." This involves familiarizing himself with the client organization, and may also involve conducting certain types of research related to other organizations doing business in the same field of endeavor. For instance, if the study area involves the shipping industry, a general indoctrination regarding this particular industry could be beneficial to the analyst. Books and reports on the subject could be useful. One of the more productive ways to meet this requirement might be to have the customer present indoctrination briefings at the beginning of the project.

It is not necessary for an analyst to be thoroughly versed in the industry field where he is to conduct his study. His training is to develop improved systems, no matter what the field of endeavor. One of the assets he brings to each new study effort is an unbiased, fresh outlook and approach.

This value of a fresh approach is not a license, however, to begin a project with no homework whatsoever. In this regard, one particular team of systems analysts will never forget (or forgive) a colleague, whom we'll call Roger. In preparation of a first assignment in the stock brokerage business, each member of the team was instructed to spend several weeks getting background material and learning the fundamentals of this area of business. Finally, the first day of the project arrived, and the customer conducted an all-day familiarization briefing for the benefit of the team. The briefing covered every facet of the stock brokerage business, including its history. The customer concluded the briefing with a statement of how delighted his firm was to have itself analyzed by a team of men of such high skill and knowledge. Then he asked the customary, "Are there any questions?"

"Yes," Roger replied. "What is a stock certificate?"

2 Develop Project Specification

At the same time the analyst is picking up system specification data during his initial contact with the potential client, certain *project* specifications must also be determined. These include a definition of the functions that are to be covered by the analysis, a determination of any possible schedule considerations, an identification of geographical locations of the functions to be studied (for the purpose of determining if there are any travel and subsistence considerations), and a documentation of whether there are any project constraints or limitations (such as related functional areas that are to be excluded from the analysis, or certain categories of data that cannot be made available to the analyst).

The Workbook, as shall be seen, provides special work sheets for developing both system and project specifications.

3 Develop Systems Analysis and Design Plan

Systems Analysis projects, whether large or small in scope, must all be planned. All the basic steps of planning must be taken. The scope of the project merely dictates the level of planning sophistication to be used. Tasks to be performed in conducting any analysis must be broken down, scheduled, and assigned. Methods for controlling the project's schedule, cost, and technical performance factors must be devised and specified. The type of systems analysis tools to be used must be decided. The selection will depend on whether the work flow of the system to be analyzed is predictable ("deterministic") or random and difficult to predict ("stochastic"). If error rates are to be checked, the approach for doing this must be chosen. If that approach involves the use of a sampling technique, the appropriate method (random, systematic, stratified, cluster, quota) must be selected. All these parameters are then incorporated into an agreed-to plan with the recipient of the analytical service.

Referring back to Figure 3, it should be noted that in the steps itemized under "System Development" there are gaps occurring at both points where project planning activities are taking place (shown under the "Project Planning and Control" category). This emphasizes another important facet of planning a project. It is not that the analysts should be unoccupied at these points, but rather should be engaged in the planning activity itself. It is an axiom of good planning that work should be planned by the key people who will be charged with the responsibility of performing the work.

4 & 5 Submittal and Approval—Statement of Objectives, and Systems Analysis and Design Plan

The two separate items that are submitted to the potential client for approval can be thought of as the "technical" and "management" sections of a traditional proposal. The technical section consists of a preliminary "Statement of System Objectives." This is a formalized summary of the basic objectives of a new or improved system. Later, as a result of detailed analysis, this preliminary statement will be expanded and refined into a more definitive "Statement of System Requirements."

The management section of what is submitted for approval at this point, consists of a breakdown of the tasks to be performed by the analysts in conducting the analysis and design effort, a schedule of when these tasks will be performed, and a delineation of what resources, in addition to manpower, will be used by the analysts in conducting the study. The plan includes a designation of assignments (which analysts will do which tasks), a technical performance plan of how the work will be directed toward trying to achieve the stated objectives, and the price or cost of performing this analysis and design work.

6 Perform Analysis of Existing System

The analysis of the existing system is often referred to as "Data Collection and System Documentation." The process involves collecting various elements of data relative to the system, then transposing this data into more workable formats either for, or as a result of, evaluation. The categories of system information involved in this transposition are listed in Figure 4, Data Categories in Terms of Prime Use.

Techniques most often used for collecting the data—the first step—are personal

DATA CATEGORIES	PRIME USE
EXISTING-SYSTEM DESCRIPTION EXISTING-SYSTEM SOFTWARE DOCUMENTATION (IF APPLICABLE) SYSTEM ENVIRONMENT FACTORS (ORGANIZATION IDENTIFICATION, HISTORICAL DATA, EXISTING HARDWARE DESCRIPTION) PROCEDURES	EXISTING-SYSTEM FLOWCHART & EVALUATION
POLICIES REGULATIONS INTERFACE REQUIREMENTS USER PREFERENCES	STATEMENT OF SYSTEM REQUIREMENTS (ORIGINAL OBJECTIVES EXPANDED & REFINED)
FORMS REPORTS INDICES CATALOGS LISTS	DATA SOURCE/USE/ELEMENT ANALYSIS
STATISTICS (VOLUME, TIME, ETC.)	COST/BENEFIT BASELINE

FIGURE 4: Data Categories in Terms of Prime Use

interviews, examination of reports and statistics, the use of questionnaires, and direct observation. Workbook checklists are used to assure that nothing is overlooked in these steps, and the information collected is inserted in the Workbook in an organized manner and recorded and summarized on work sheets.

As to interviews, it is through this means that the systems analyst establishes his personal relationships that are so important in any system study. There are both positive and negative values involved in interviews. On one hand there are those persons who are comfortable and secure in their present environment and fear and resist the prospect of change. On the other hand, the very people who perform the work are often the best sources of suggestions for improving the system.

If a project is very large, a questionnaire form may prove to be a practical approach for gathering certain categories of data. If this is the case, care must be taken in developing the questionnaire. Questions must be succinct and to the point so they elicit a response that is useful. Furthermore, the questionnaire should be easy for the questionee to understand and answer.

Direct observation of the system in operation is another of the techniques used. Again, data derived in this manner is recorded on the appropriate work sheets in the Workbook. Sometimes in this approach the analyst will add a "diary" section to the Workbook, recording all data pertaining to the system flow in narrative form.

Factors relative to each of the "use" categories shown in Figure 4 are described more fully as follows:

Existing-System Flowchart and Evaluation. One of the most useful results of the data collection activity is the development of a flowchart of the existing system. In developing this system flowchart the analysts learn the details of the existing system more effectively than if a flowchart had been provided them by the customer. In the process of developing the flowchart the analysts begin to recognize the potential changes that would lead to an improved system. It is true that the solution to a problem is inherent in the understanding of the problem.

There are several classes of flowcharts used in recording study data in the Workbook. The purpose of any chart, of course, is to clarify and to make the information more understandable. One of these types of charts is a process flowchart. It concerns itself with the flow of physical materials, including documents, through a system, especially in terms of distance and time. It is most useful in analyzing some of the cost and benefit factors for existing and proposed systems.

System flowcharts, which are described in detail in a later chapter, have been called the analyst's "shorthand." They can be forms-oriented or task-oriented. These flowcharts are not only the primary way of recording data pertinent to the current system, but are used for developing and displaying the new system as well. Later, in the implementation phase, *program* flowcharts, a fundamental tool of programming, would be developed.

There are several types of input necessary for constructing the existing-system flowchart. One category of information, of course, involves a description of the existing system derived from interviews, observation, and an examination of procedures.

If the current system is already "computerized," documentation relative to the existing system's "software" (data processing programs) can also prove useful. In one actual case involving a systems analysis of a stock exchange clearing function, one of the main problems involved the number of "fails" occurring in the process. The analysts obtained the documentation of the current system and examined it in their hotel room that evening. They found that by making a minor change in a mathematical procedure, they could reduce the number of fails by 25%. Furthermore, the amount of effort involved in making this change required only three days of programming time. This became one of their significant "short range" recommended improvements for the system.

Information regarding the current system's environment also serves as necessary input to the development of the flowchart. This information includes a more detailed description of existing hardware (than what was collected earlier in the planning stage), and other similar types of data. Historical information can prove useful. It could be important to know what has happened in the past on the system. It might be, for instance, that the volumes of data increased significantly in the recent past. Information as to changes in organizations, functions, and responsibilities could also prove useful. Then there should be data collected that defines and describes the organization or organizations involved in the study.

Statement of System Requirements. One of the most important elements of this part of the analysis of the existing system is the definition of detailed system requirements. In this effort the orginal Statement of System Objectives is expanded and defined in greater detail. Detailed objectives and constraints in the customer's environment should, by this time, have been identified in detail.

A determination and definition of interfaces with any other existing system, and requirements within or between departments, should have also been completed. This involves all organizations that either are sources of data or that require data from this particular system. System features that the system users would like to have incorporated in the new system would have also been recorded. In other words, at this point in the analysis, problems and requirements should have been defined to the extent that a more detailed goal of desired system characteristics can be formulated. Once reviewed and approved by the client, this Statement of System Requirements is used as the baseline, or "specification" for the design of the new or improved system.

Data Source/Use/Element Analysis. No analysis would be complete without either gathering the actual records and documents relative to the system, or at least recording in the Workbook the essential information from this data, including data elements, when necessary. Information categories include reports, forms, indices, catalogs, and lists. The analyst should try to obtain a copy of every document used in the system.

One of the steps in the process of evaluation is the analysis of these reports, forms, and other documents. These items are studied from the standpoint of what they contain, what their purpose and use is, who prepares them and how, and how often they're issued. The source of the data is reviewed and the distribution and number of copies produced is questioned.

Cost/Benefit Baseline. During the course of the study, and interrelated with the previously described data collection steps, a continuing evaluation of the existing system should take place. The evaluation should cover system efficiency, accuracy, timeliness, economy, and productivity, often referred to as "cost and benefit factors." Various system simulation techniques might be used to accomplish these evaluations.

All types of statistics and other data from which system cost can be calculated and system effectiveness measured must be collected. This includes processing times, volume of work, number of people involved, salary or rate ranges, and other similar data.

Costs are calculated for all elements of the system. Costs are divided into recurring costs, such as salaries, and non-recurring costs. Efficiency is measured and system capabilities are analyzed. All elements of the existing system are examined from the standpoint of what steps could be eliminated, combined, changed, simplified or improved. All of these cost/benefit factors as they relate to the existing system are assembled into a "baseline" against which later proposed system improvements can be measured.

7 Control Project (Analysis Tasks)

Throughout this analysis phase of the project it is controlled as to tasks being performed, schedule compliance, budget compliance, and technical performance. The level of sophistication used depends, as it did with planning the project, on the scope and complexity of the project. If the project warrants it, there could be periodic review meetings with the client and written reports as well. As will later be shown, the Workbook serves as a useful, time-saving tool in this regard.

8 Interim Approval—Statement of System Requirements

The Statement of System Requirements, the updated, more defined version of the originally agreed-to Statement of System Objectives, can now be reviewed, changed if necessary, and approved by the client. Again, this is one of the important steps to make sure the analysis proceeds in the direction of producing a system that is close to what the client needs and wishes to have.

There are other items that the client can also review at this time. One is the flowchart of the existing system which can be checked from the standpoint of accuracy. Another item that can be productively reviewed is the list of preferred system characteristics which the client's personnel asked be incorporated in the design of the new system. Management may or may not concur with some of these requests and, at this point, can reject undesired requests, modify them, or add to the list.

9 Design New System

The system design, as previously mentioned, gradually evolves in the mind of the analyst as the existing system is analyzed, questioned, and evaluated. Just as there was a system flowchart for the existing system, the analyst now prepares a flowchart for the new system concept. Possibly several alternate approaches may be prepared in flowchart form so as to compare them from the standpoint of best cost and benefit factors. If the system design involves new or unique data processing techniques, a preliminary *program* flowchart of the design concept might also be prepared at this time.

Mock-ups of new forms, reports, and other system documents would also be prepared. These would be "keyed" to the new-system flowchart.

As a means of communicating to the customer that requirements have, indeed, been met, the flowchart should be cross-referenced with each numbered paragraph, or "statement," of the Statement of System Requirements. Examples of how this is done in the Workbook are shown in a later chapter.

Costs of operating the new system and benefits derived from its operation should be determined. These figures should then be compared with the existing system's cost/benefit baseline.

In developing the design of a new system, alternate approaches can be evaluated using various available tools: Linear Programming can be used to determine the best mixture of components; Queuing Theory can be used to predict the workload on any single segment of the system at a given point in time; Forecasting can be used to predict future workloads based on past performance and known variables; Simulation and Mathematical Modeling can be used to examine the effect of changes to a system.

If it is determined that existing data processing equipment would not be sufficient for the new system concept, there might be the need for a feasibility study. This study would be for the purpose of determining what equipment configuration would be best for the new system. In the feasibility study the analyst would concern himself with the equipment description and cost, including a cost comparison as to the option between purchasing and leasing the equipment. The analyst would also determine delivery restraints, if any, and the availability of maintenance support for a new system. Requirements for new or modified facilities for the hardware would also be covered in the study. The analyst would also determine what programming support was available from the manufacturer, and the extent of rewriting (or conversion) that would be necessary relative to existing programs.

10 Control Project (Design Tasks)

Following the sequence of steps illustrated back in Figure 3, just as there was project control exercised over the analysis tasks, the same process of controlling schedule, cost, and technical performance factors would take place as the system is being designed.

11 Develop Implementation Plan

Once a new or improved system has been developed, as shown in the sequence of steps illustrated in Figure 3, a plan for implementing the system must be devised. This plan, like all project plans, is a matter of breaking down the tasks necessary for implementing the system as a first step, finding their inter-relationships and scheduling them, developing a plan for measuring technical performance during implementation, determining what resources are required for implementation, grouping and assigning the tasks, and pricing them. A description of how the project would be controlled during implementation should also be devised.

In the breakdown of tasks there are several functions that would probably be included in a computer software implementation plan. These would include the preparation of written instructions, training of personnel in the operation and use of the new system, and obtaining the equipment (if new equipment was specified) and the forms and other supplies needed for the new system. There would also be the need for converting existing records and files to the new configuration and, possibly, the creation of new files. Finally, there would be the programming required for the new application.

12 & 13 Submittal and Approval—System Design, and Implementation Plan

Phase I (Systems Analysis and Design) concludes with the submittal and review of the recommendations, and the decision as to the next course of action. Often the recommendations and plan are submitted in the form of a "final report" derived from key sections of the Workbook. This report would contain two main sections. One would be a description and report of the recommended design, along with operating-cost information. The other section would contain the project plan for implementing the system.

14 System Programming and Implementation

This part of the overall process can be roughly divided into three categories of work. The first is the development of detailed specifications. Once the system concept has been approved by the client, it is now appropriate to define the computer system in detail, including the definition of the data fields. There should also be a detailed definition of the system paper content, and specifications developed for the system documentation if this latter area has not been previously agreed to.

The second category of work involves procedurizing and programming the system. This also involves unit and system testing, and the development of a test data base and sample input data in support of the tests. Also involved are the tasks preparatory to installing the system, including converting files and establishing the system data base.

The third category of work involves installing the system on the client's equipment (or client-designated equipment) and performing pilot operations. These are often run in parallel with the old system. When the new system is proven, it becomes operational and is turned over to the client.

Concurrent with these three categories of work several other tasks may also be performed, if applicable. If the new system design involves new hardware and new, or modified facilities, tasks involving the ordering, delivery, installation, and test of these materials takes place. If training of client-personnel in the new system operation and use is desirable, there could also be tasks involving the development of training aids and the conducting of training sessions.

15 Control Project (Implementation Tasks)

As with the other previously described steps in a systems analysis project, the performance of the implementation tasks are not left to chance, either. Schedule, cost, and technical performance factors must be controlled, that part of technical performance control involving unit and system test being of extreme importance.

16 New System Operation

Once the new system has been installed and successfully operated as a pilot system, it is turned over to the user. It is now an operational system and the systems analyst's task is completed.

CONSTRUCTING THE WORKBOOK

The Workbook provides work sheets and checklists to assist the analyst in all the foregoing steps of systems analysis. The work sheets are used to record information in an organized manner, as it is collected or developed, so that it will be useful and easily accessible. Each work sheet includes space for recording logically related elements of information, such as the data needed to identify and describe the organization being studied, or the set of interfaces with outside organizations that occur for a particular department, or the data on existing computer "software" used. The checklists are used in planning a particular step, and as checks or reminders to insure against leaving out a task or neglecting to develop a needed piece of information.

As information is developed and the work sheets filled in, the Workbook becomes an evolving catalog of information elements useful to either an individual analyst or an entire team as a system is analyzed and a new or improved system of operations is designed. It also serves as a constantly up-to-date record of progress throughout the systems analysis and indicates, by readily observable blank spaces, the tasks remaining to be done, or the information yet to be developed or collected.

A complete set of work sheets and checklists is described in this book. Together, they make up a Workbook that should be considered as general purpose, useful in many systems analysis projects. There will be some instances, however, where the Workbook will be more useful if it is "tailored" for a specific project. In such cases the work sheets and checklists described here can be productively used as guidelines in the development of unique or special purpose forms.

In most studies, regardless, the major categories of data are those represented by the "tabs" of the Workbook shown in Figure 5. This example of the Workbook provides separate sections for work sheets and checklists relative to:

- Developing the original analysis and design plan;
- Identification of various system equipment factors (such as organization units);
- Describing the existing-system flow;
- Describing the existing-system documents;
- Describing requirements, including interfaces, regulations, and other legal or policy data affecting the systems;
- Recording evaluation data, including information about economy, efficiency, accuracy, timeliness, and productivity of the existing and new systems;
- Developing and describing the new or improved system;
- Devising and describing the implementation plan, documenting the system, and monitoring and evaluating its operation.

There is also a glossary, as the first part of the Appendix section, for recording special terms and their definitions used in the area under study. Many businesses and professions (including the computer field) have their own "Towers of Babel" consisting of common terms that have different meanings, or terms, codes, or abbreviations that are unique to one field of endeavor. The Appendix section can also be used for holding, in one place, the miscellaneous material that analysts invariably pick up during the course of a study.

When using the Workbook, the work sheets should be filled out by the analyst and not by the interviewee. Often, in actual use, the analyst will merely jot down quick notes on the Workbook forms during the interview. Later, after the interview has been completed, he will complete the forms in more detail.

Workbooks should be bound in loose-leaf binders. This will allow the systems analyst to replace the work sheets he has used for note-taking with work sheets which are more legible, or with better organized information when he has transcribed his data. It also facilitates the addition of extra or new forms, special notes, copies of reports, exhibits, and other project data. If it is decided to "tailor" a workbook for a unique project, the loose-leaf approach allows use of existing work sheets interspersed with whatever special purpose work sheets have been developed.

If there is more than one systems analyst on a project, each should have his own Workbook. The individual analyst's Workbook would contain only those work sheets and

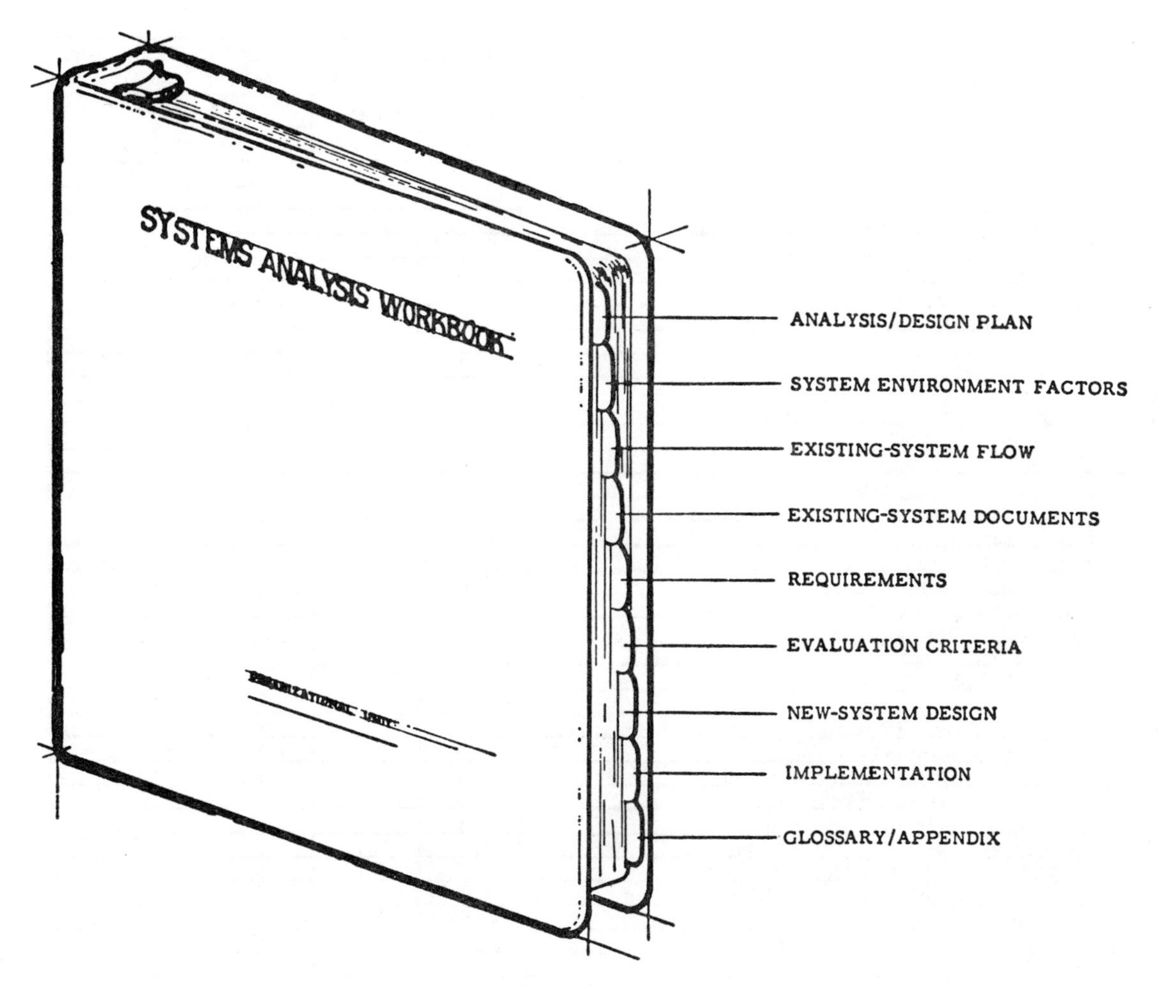

FIGURE 5: Systems Analysis Workbook

checklists pertaining to the particular area he is to study. This reduces the bulk of each book and allows the individual analyst to see his own progress and the tasks ahead more clearly. As the project advances and it becomes important to consolidate all the information, individual Workbooks can quickly and easily be assembled into a master Workbook.

The first section of the Workbook, as previously shown, is the Analysis/Design Plan section. First in this section is the Workbook Contents Checklist, a two-page form illustrated in Figures 6A and 6B. This two-page checklist lists available pages in the Workbook and is designed so as to assist in assembling one or more Workbooks for a specific project. It can also serve as a table of contents for individual Workbooks. In use, the need for specific existing work sheets or checklists is indicated with a check mark. If more than one particular type is needed (such as is often the case with the Sub-Unit Identification work sheet, for instance), the quantity can be listed alongside the check mark.

1.1A

WORKBOOK CONTENTS

Checklist

Sheet 1 of 2

	Check Sheets Needed For Workbook Indicate Quantity If More Than One (1)						
	Sheet No.	Master Copy	Analyst				
			A	B	C	D	E
TAB:ANALYSIS/DESIGN PLAN							
Workbook Contents Checklist	1.1						
Project Specification-Scope/Limitations	1.2						
Project Specification-Schedule & Location	1.3						
System Specification	1.4						
Objectives Checklist	1.5						
Task Planning	1.6						
Resource Estimating	1.7						
Task Assignment	1.8						
Design Review Checklist	1.9						
Interview Planning Checklist	1.10						
TAB:SYSTEM ENVIRONMENT FACTORS							
Organization Identification	2.1						
Sub-Unit Identification	2.2						
Key Personnel	2.3						
Headcount	2.4						
Plans and Trends	2.5						
Historical Data	2.6						
Equipment Description-Existing DP	2.7						
Equipment Description-Data Comm.	2.8						
Equipment Description-DP on Order	2.9						
TAB:EXISTING-SYSTEM FLOW							
System Flow	3.1						
Features Susceptible to Improvement	3.2						
Procedure Summary	3.3						
Automated System Identification	3.4						
Automated System Breakdown	3.5						
Automated System Summary	3.6						
Developmental Program Status	3.7						
TAB:EXISTING-SYSTEM DOCUMENTS							
Existing-System Document Identification	4.1						
Existing-System Manual File Description	4.2						
Existing-System Data Base Description	4.3						
Data Element Inventory	4.4						
Data Element Matrix	4.5						
TAB:REQUIREMENTS							
Intra-Organization Interfaces	5.1						
Inter-Organization Interfaces	5.2						
External Interfaces	5.3						
Regulations	5.4						
Policies	5.5						
User Requirements & Preferences	5.6						
User Preferences Checklist	5.7						
TAB:EVALUATION CRITERIA							
Existing-System Operating Costs	6.1						
Existing-System Benefits Data	6.2						
Existing-System Value Meas.-Economy	6.3						
Existing-System Value Meas.-Efficiency	6.4						
Existing-System Value Meas.-Quality	6.5						
Existing-System Value Meas.-Accuracy	6.6						
Existing-System Value Meas.-Reliability	6.7						
Operating Tasks Breakdown	6.8						
Resource Estimating	6.9						
Operating Cost Comparison	6.10						
Cost/Benefit Comparison Summary	6.11						

FIGURE 6A: Workbook Contents Checklist—Sheet 1 of 2

Checklist Sheet 2 of 2

	Check Sheets Needed For Workbook Indicate Quantity If More Than One (1)						
	Sheet No.	Master Copy	Analyst				
			A	B	C	D	E
Life-Cycle Cost Comparison	6.12						
Impact on Profit and Cash	6.13						
Existing-System Cost Breakdown	6.14						
Proposed-System Cost Breakdown	6.15						
TAB:NEW SYSTEM DESIGN							
New-System Synopsis	7.1						
New-System Flow	7.2						
New-System Features Summary	7.3						
New-System Environment Factors	7.4						
Manual Operations Procedure Requirement	7.5						
New-System Input/Output Synopsis	7.6						
New-System Document Identification	7.7						
New-System Data Base Description	7.8						
New-System Manual File Description	7.9						
Decision Table	7.10						
Equipment Requirements and Evaluation	7.11						
Equipment Utilization Summary	7.12						
New-System Equipment Configuration	7.13						
Site Preparation & Activation Consideration	7.14						
Forms Analysis (Sheet 1 of 2)	7.15						
Forms Analysis (Sheet 2 of 2)	7.16						
Data Element Design	7.17						
Distributed Processing System Equipment	7.18						
Small Computers-Comparison	7.19						
TAB:IMPLEMENTATION							
Implement. Project Specification Plan	8.1						
Implement. Project Specification Files	8.2						
Implement. Project Specification Equipment	8.3						
Implement. Project Specification Schedule	8.4						
Task Planning	8.5						
Resource Estimating	8.6						
Implement. Cost/Price Summary	8.7						
Orientation Requirements	8.8						
Skills Inventory	8.9						
Skills Requirements	8.10						
Training Requirements	8.11						
Technical Performance Review Materials	8.12						
Task Assignment Matrix	8.13						
Task Assignment	8.14						
System Design Material	8.15						
Plan & Control Material	8.16						
Contract Considerations Checklist	8.17						
Presentation Considerations	8.18						
Presentation Audience	8.19						
Presentation Planning	8.20						
Procedure Planning	8.21						
Operating Procedure	8.22						
Program Summary	8.23						
System/Program Spec. Manual Checklist	8.24						
Operating Instructions Manual Checklist	8.25						
System Evaluation Factors Checklist	8.26						
Time Evaluation	8.27						
Cost Evaluation	8.28						
TAB:GLOSSARY/APPENDIX							
Glossary	9.1						

FIGURE 6B: Workbook Contents Checklist—Sheet 2 of 2

IDENTIFICATION AND SEQUENCE OF WORK SHEETS AND CHECKLISTS

Complete lists of the Workbook's work sheets and checklists appear in two places in this book: Figure 6 (2 sheets), and in Chapter 21, Recapitulation. The chapters between present detailed explanations of the application and use of these materials. As presented in these chapters, though, some selections (in order to provide a more coherent explanation) were discussed in a different sequence than would be used in assembling them into a workbook. For this reason, the Chapter 21 list cross-references the work sheets and checklists to the book's figure numbers.

As a further help to the reader, the work sheets and checklists, in addition to an identification as such in their titles, can further be easily identified as distinct from other illustrations through the decimal numbers that appear in the upper right-hand corner of each such sheet. These numbers correspond with those that appear on the two previously mentioned lists.

2
Starting Out Right—Defining the Project

Success in a systems analysis project, as it is with most any endeavor, depends on proper and thorough preparation. The detailed steps required to prepare for one of these types of projects involves:

- Defining Project Specifications
- Defining System Specifications
- Becoming Familiar with the Business Environment
- Adapting the Workbook
- Planning the Project

The first steps of defining project and system specifications consist of initially communicating with the client of the analytical service (this could be an in-house department), and defining the study's problems, objectives, and constraints. Schedule considerations and study-location factors must also be determined. Then there are several categories of familiarization that must be considered. General familiarization with the client organization, one of the prime purposes of the initial contact with the client, is of vital importance, of course. In addition to this, though, are circumstances where it is of value to become familiar with the general business environment involving the client's field of endeavor and, in some cases, to construct a glossary of special words or terms that may have unique significance within that field.

With this foregoing data in hand, the workbook can be adapted or modified, if necessary, to meet any special conditions of the study, and the analysis/design plan can be formulated. This chapter deals with all these necessary steps involved in preparing for a systems analysis project.

DEFINITION OF SYSTEM AND PROJECT SPECIFICATIONS

The first section of the Workbook, Analysis/Design Plan, contains work sheets and checklists useful in preparing for a systems analysis project. These aids are sequenced in the order in which they most likely would be used.

The Specification Work Sheets

Two Project Specification work sheets and a System Specification work sheet are the first fact-gathering aids in the Workbook. They are to be filled out with the data resulting from the initial contact with the client. Information to be gathered and recorded on these work sheets provides the essential input required to formulate a detailed project plan.

The heading data on the first of two Project Specification work sheets, Figure 7, describes the general area to be studied, and provides space for recording the name, address, and phone number of the specific client-individual who is to be considered the analysts' chief contact for the duration of the project.

The main body of the form provides space for listing specific functions to be covered in the analysis. Space is also provided for describing specific project constraints or limitations.

The Second Project Specification work sheet, Figure 8, provides space for briefly citing other data important to the development of the project plan. This includes schedule considerations, study area locations (so that travel and subsistence factors can be determined and planned for, if they apply), and an often-overlooked area of concern regarding whether or not work space for the analysts will be provided at the client's facility. (If not, other provisions must be planned for.)

The System Specification work sheet, Figure 9, identifies the name of the system involved in the analysis, and provides space for listing problems being experienced with the current system, and prime objectives of a new or improved system. In this latter category, a checklist of potential objectives can be used. Such a checklist is described later in this chapter.

The System Specification work sheet also provides space for describing specific *system* constraints or limitations, and space for briefly describing existing data processing and/or communications equipment used in the current system. This type of information can assist in determining what types of personnel should be assigned to the project. If the software for the current system is documented, and if that documentation is available, both facts are also noted on the work sheet.

Identification of Problems, Objectives, and Constraints

In filling out the specification work sheets, it is important to keep in mind that the essential requisite to planning is to establish firmly a definition of the problems with the existing system, the objectives of the project (the purpose of the analysis and design effort), and the system and project constraints (the limits or boundaries). That is, defining what can and cannot be done. This subject is treated separately here for two reasons. First, these factors are often taken for granted when they should be formally defined and

PROJECT SPECIFICATION

Scope & Limitations

General Area(s) to be Studied __________

____

Chief Contact (Name) __________

Address __________ Phone __________

List Specific Functions to be Covered in Analysis: __________

Describe Specific Project Constraints/Limitations: __________

FIGURE 7: Project Specification—Scope & Limitations

PROJECT SPECIFICATION

Schedule & Location Considerations

Schedule Factors:

Project-Start Considerations____________________

Project-Complete Considerations____________________

Study Area Geographical Location(s)____________________

If Travel & Subsistence Required, Specify____________________

Will Client Provide Work Space, if Required?____________________

Describe____________________

FIGURE 8: Project Specification—Schedule and Location Considerations

SYSTEM SPECIFICATION

Name of System ____________________

____

List Current Problem(s)

____
____
____
____
____
____
____
____

List Prime Objective(s)/Goal(s)--(See Checklist)

____
____
____
____
____
____
____
____
____

Describe Specific System Constraints/ Limitations ____________________

____
____
____
____
____

Briefly Describe Existing Data Processing and/or Communications Equipment ____________________

____
____
____
____
____

Is Current System Documented? ____________

Is Documentation Available? ____________

FIGURE 9: System Specification

agreed to by all concerned. Secondly, they are essential to a proper plan, and without a plan, a project cannot be truly managed.

It is usually because the client recognizes that a problem exists that he engages the services of a systems analyst. It is, furthermore, only a semantic exercise to twist the statement of a problem around into a statement of an objective. Of prime consideration, though, is the fact that with problems the difficulty of "stated" and "discovered" problems is encountered. All that can be dealt with at this point in the analysis preparation process is the stated, or known, problem or problems. It should be understood that it is likely that additional problems will be discovered during the course of the analysis. This is one of the main reasons that the *detailed* statement of system requirements is formulated later, as a result of detailed analysis, rather than at the beginning of the project.

A proper preliminary objective for a project would be a continuing improved state or condition which should be achieved when the new system is successfully implemented. This condition might be a certain level of revenue or profit, it may be a specified normal turn-around time for reports or for responding to customer orders, or it may be a reduction in manhours required for a particular task or set of tasks. Rarely will an objective for system design and implementation be a unique or "one-time" event.

Objectives recorded on the System Specification work sheet, even though preliminary in nature, should be specific. It is never sufficient to state an objective in terms of simply improving an existing system or of implementing a computerized system. The idea that a new system or an "automated" system is a better system has been a popular concept too long. An improved system, per se, is of no benefit to a business client; implementing a better system in order to increase profits or reduce costs is of great benefit.

Objectives can generally be classified into one of a very few categories. As shall be seen later, these relate closely to the criteria for evaluating proposed improvements. The list that follows may not be all inclusive, but will serve as a guide to the type of objectives to look for.

Reduced Waste. This includes such goals as decreasing the workload of an over-worked department or group, reducing overtime required to accomplish a repetitive task on schedule, freeing skilled people from routine work so they can perform tasks more critical to the main operation, and eliminating the need for "a man and a half" to perform a particular job. Related improvement criterion: Efficiency.

Reduced Cost. Cost reduction is probably the most frequently stated objective. In most businesses and in goverment agencies as well, all things are measured in dollars. No matter what the stated objective is it can, and will, be translated into cost saving (or the other side of the coin, increased profits). Often cost reduction, per se, is a stated objective. Even if it is not, the analyst should be prepared to measure success in these terms. Related improvement criterion: Economy.

Better Information. Better information may mean more details or it may mean more accurate data. This is a more complex problem than it might appear since diminished accuracy or lack of detail may result from inadequacies in the raw source data, or from the computations and manipulations applied to it. It may also result from shortcomings in the format of the output. Finally, accuracy may be intimately related to the timeliness of the data, a problem that rates a category of its own. Related improvement criterion: Accuracy.

More Current Information. This also has a dual meaning. It might relate to a more rapid response time for requested data, usually data wanted for a special report or to solve an acute or urgent problem. However, the more significant system objective usually relates to the currency of periodic outputs. In some organizations a recap of sales or expenses, for example, can "wait until Friday" to be made available; even monthly summaries are adequate for some. On the other hand, an associate of the authors at one time developed a system for a *daily* corporate projected profit and loss statement which was successfully implemented, solving a specific problem in a major transportation firm. Currency, then, is relative. In all control systems, though, the data should be current enough so that if problems are disclosed there is enough time in which constructive corrective action can be taken. Related improvement criterion: Timeliness.

More Output from Available Resources. Increasing output is the final general category of objectives in this list. This is a highly touted advantage of automation although not all computer applications successfully meet the objective. Resources may include dollars (translated into budget figures or allocations), manpower, or equipment. More often than not the objective is to increase productivity by using the same level or reducing the combination of all these resources. This type of objective differs subtly from cost savings and reduction of waste. Related improvement criterion: Productivity.

The Objective Considerations Checklist

The list given above is summarized into an Objective Considerations Checklist, Figure 10, and should be used in conjunction with the System Specification work sheet. The checklist, itself, deals only with general categories and sub-listings of objectives. Unless these objectives are defined more specifically on the work sheet they will be of little value in planning the project. Objectives for a particular project must be specific. If they fall into the category of Reduced Cost, the questions that must be answered are: What costs? In what department? For what type of tasks? An example of a specific objective in the category of reducing waste might be:

> To eliminate redundant files and reduce the amount of clerical effort expended by division heads and directors in monitoring the status of research grants.

The categories are also inter-related. Productivity, as has been pointed out, relates closely to economy and efficiency. Accuracy relates closely to timeliness. Other relationships are also apparent. Furthermore, once an objective is stated formally and precisely, it may overlap several of these categories. For example, a management system may be requested to increase the effectiveness of department heads in controlling their budgets by providing weekly instead of monthly expense data for their departments. This involves timeliness, increased productivity, and better information. Other terms that can be used when considering system objectives include profitability, reliability, security, safety, quality, flexibility, capacity, acceptance, usability, and adaptability.

If there is more than one objective, as there usually is, they must be ranked or given a priority in terms of importance to the client. The ranking of objectives will later determine the priority or weight to be given to the various evaluation criteria. It will often be a guide to selection of two conflicting system alternatives, one of which, for example, might reduce operating costs by reducing the amount of computer core required for a set of programs and thus allow a smaller computer to be used. The alternative might increase

OBJECTIVE CONSIDERATIONS

Checklist

	Check
REDUCE WASTE (Efficiency)	
Decrease Workload	
Reduce Overtime	
Free Skilled People from Routine Work	
Reduce "Down" or "Off" Time of Equipment	
Increase Off-Line Operations	
REDUCE COST (Economy)	
Reduce Number of Operations Performed	
Reduce Size or Quantity of Equipment Needed	
Reduce Manhour Requirements	
Reduce Number of Reports Produced	
Reduce Distribution of Reports	
BETTER INFORMATION (Accuracy)	
Increase Accuracy of Input/Output	
Select Better Reporting Elements	
Improve Distribution (Right Data to Right People)	
Increase Precision or Capacity of Equipment Used	
Increase Flexibility or Variety of Processing Operations	
MORE CURRENT INFORMATION (Timeliness)	
Decrease Throughput Time (Effective Date of Information)	
Cut-Down Process Turnaround Time	
Reduce Reproduction Service Time	
Reduce Distribution/Transmission Times	
Provide More Frequent Reports	
MORE OUTPUT FROM AVAILABLE RESOURCES (Productivity)	
Increase Hardware Utilization	
Provide More Compact or Efficient Storage	
Provide Greater Compaction of Data	
Eliminate Redundant Operations	
Eliminate Unnecessary Operations	

FIGURE 10: Objective Considerations Checklist

accuracy and detail of information but at the expense of greater core requirements. The two are in direct conflict and only by establishing priorities can a proper choice be made.

Priorities and ranking of objectives must also have the concurrence of the client for reasons similar to those explained relative to the choice of objectives.

Once objectives have been determined, project constraints and limitations must be determined. They might apply to the system itself, or to the conduct of the project. System constraints might involve a requirement that only existing computer and peripheral equipment can be considered for use in the design of the new or improved system. Such stipulations as to the use of available resources are common. Project constraints are usually in terms of organizational boundaries, functional areas, perhaps geography, management levels, or types of data.

The analyst may, for example, be limited to studying marketing and sales functions, only, or activities relating to production. On the other hand the analyst might be studying only the marketing department but not be limited to marketing activities. That is, the analyst would be looking not only at marketing procedures and data, but personnel functions and administration within the marketing department as well.

It is not surprising to occasionally run into special client-imposed constraints that can only be classified as remarkable. In one actual instance, the general manager of a fairly large manufacturing firm requested the consulting analysts to design a new automated management reporting system, but with the limitation that the analysts were not to have access to the firm's own data processing department. It seems that the data processing department's manager had failed in his own attempt to develop the reporting system, and the general manager was reluctant to embarrass him.

Occasionally limits will be expressed in terms of the types of reports and information required. For example, the analyst may be limited to fiscal information as his only concern, or be limited to inventory data, or purchasing data.

Once system problems, objectives, and constraint factors have been determined, the basic goals of the analysis should be summarized and ranked in order of priority on a "Statement of System Objectives." This, together with a project plan (described later in this chapter) are then submitted to the client for approval.

PRE-ANALYSIS FAMILIARIZATION

No matter who is doing the analysis, some familiarization is necessary. For the inside staff man not as much will be needed as for the outside consultant. However, even for members of the organization, much that is "known" will be found to be based on assumption, hearsay, or a biased viewpoint. Therefore, it is well to examine systematically and question what is "known" and add to this knowledge whenever necessary.

The Client Organization

During the initial contact phase, it is well to find out as much as possible about the specific organization to be studied prior to planning the project in detail. It is also useful to do some outside research in this area as well. If the company is a large one, such standard sources as Moody's Manuals, Standard and Poor's Register, and Dun and

Bradstreet's various directories can be checked. These sources will not only provide a list of the principal officers and managers, but a brief history of the company and a description of their products or services, as well. Smaller firms may be described in one of the regional or specialized guides such as the Dun and Bradstreet Reference Book.

If the area to be studied is a Federal Agency, the U.S. Government Organization Manual can be useful in this research. It provides lists of officials and descriptions of the mission and function of each agency, its subordinate units and parent organization, and describes groups on the same organization level, as well. Some states and municipalities publish similar documents which are of great value in studying state and local government agencies.

Most major government agencies publish annual reports of progress for the legislative branch; these are of value also. Annual reports from private corporations, and general promotional brochures are other possible sources of useful background material.

In short, background material regarding a specific client can prove useful to the analyst and assist him in doing a better job for his customer. It's worth the effort.

The Business Environment

For the outside consultant who is not a specialist in the particular type of business being studied, a certain amount of research will be essential. It is not necessary to become a specialist but, as was illustrated in the preceding chapter, it is important to be able to talk intelligently about the basics of the business. In the shipping/transportation field, for instance, there has been a trend in recent years toward "intermodal transportation systems" that allow a shipper to use one bill-of-lading regardless of the types or number of modes of transportation required to get a given shipment of goods to its destination. If this is to be the subject of the systems analysis project, then it would be well for the analyst to become familiar with what others in this industry have been doing in this regard. Information as to approaches, general problems, regulatory agencies, typical organization structures, and other similar data can prove useful.

The history of the business field can be helpful, especially recent significant events. Some activities will have resulted in problems which are common to all similar firms; knowledge of them may help the analyst better understand specific problems which may come up during the project.

This type of familiarization need not be an unpleasant task, nor should it take a great deal of time. Several hours in a good library and some light reading may be sufficient. In a field that is completely unfamiliar, a good encyclopedia could be a starting point. Possibly textbooks may exist for subjects related to the field under study, or technical or business books on the subject may exist. Scanning a few recent issues of the appropriate trade journals is a must, if they are available. Ads in these magazines should not be overlooked. The analyst will be surprised how rapidly he can get a feel for the current concerns in an industry by studying its journals and news magazines.

Jargon

Many fields have their own special languages, or "jargon." In this pre-analysis familiarization phase, the analyst should try, in his research, to become familiar with this

9.1

GLOSSARY

Terms, Abbreviations, Codes

Expression	Description
ABF(I)	AM BID FIRM (IMMEDIATE)
AINOK	ADVISE IF NOT OK
AOAP	ALL OR ANY PART
AON	ALL OR NONE
BMF(I)	BID ME FIRM (IMMEDIATE)
CXL	CANCEL
DD	DELAYED DELIVERY
FADED	PREVIOUSLY QUOTED BIDS WITH-DRAWN OR REVISED DOWNWARD
GIVE UP	THIRD PARTY "GIVES UP" NAMES OF BUYER AND SELLER SO THEY CAN DEAL DIRECTLY WITH EACH OTHER

FIGURE 11: Example of Glossary of Terms

language. It could be useful for the analyst to construct a "Glossary of Terms" (the first section of the Workbook's appendix section), such as the example (for the stock brokerage business) shown in Figure 11.

A particular benefit of reading the trade journals is that each field's special terms are used frequently in the articles and advertisements. The analyst will find the most common and most current terms used because the journal writers usually make a particular effort to keep up with this aspect of their field. Often, current business or technical books on the subject-field will contain useful glossaries.

The analyst should make notes as to unfamiliar words, terms, and abbreviations while doing his research. Definitions may be implied by their context, or can be looked up in reference material. Terms whose definitions cannot be satisfactorily determined in advance can later be reviewed with someone in the client organization. Devoting a short time at the beginning to purposefully clarifying terms can greatly enhance the analyst's efficiency throughout the period of the study effort.

ADAPTING THE WORKBOOK

Having become familiar with the business environment and the client organization, and having established project and system specifications, the analyst should now review the Workbook. The purpose is to adapt or modify it, if necessary, to meet any special project or system conditions. There should be no hesitancy to make changes, but care must be taken not to make the Workbook more complex. If it is found that the basic structure is adequate but a greater level of detail is called for, instead of adding more spaces to the work sheets, it might be more useful to add checklists instead. Whatever changes are made, it should always be kept in mind that the work sheets or checklists developed for a given project should be designed in such a manner as to allow flexibility in the interview and flexibility in recording the information.

In adapting or modifying the Workbook consider the deletion of items on the standard work sheets, the redistribution of space to shift emphasis, and the substitution of whole pages in those cases where this might be desirable. For example, if no automated system currently exists, the Workbook pages relating to current software will be superfluous. However, under these circumstances it might be that more space will be needed, and perhaps an additional special page or two, for recording more detailed data on manual procedures used.

The last chapter of this book contains a complete listing of all work sheets and checklists. Preceding this list is a brief description of a sample systems analysis project. The materials that were selected for use in that project are indicated, by asterisk, on the list. An examination of that example might provide a further understanding of how work sheets and checklists may be chosen for specific applications.

PLANNING THE PROJECT

Project management is the process by which it is assured that the objective is achieved and resources are not wasted. Planning is one of the two essential parts of project management. Control is the other.

Probably the most neglected area in systems analysis involves the planning and control of the project, especially those projects requiring automation. More than one disastrous project has been launched by "computer people" who communicated their aims to the vexed manager using technical data processing jargon in lieu of specific lists of easily understood tasks, schedules, and costs. This problem applies equally to in-house projects or those requiring the services of outside consultants.

Each project must first be planned in detail. Control is involved with comparing actual progress with the plan and taking corrective action when the two do not correspond. Without the plan, true control is not possible; the need for corrective action, its nature, extent, and urgency cannot be accurately determined.

No matter what the size or scope of the project is, it should be planned and controlled. This does not mean, however, that the same level of planning effort should go into projects, regardless of size. One particular systems analysis project that was in the neighborhood of $8,000 in cost involved the planning effort of three systems analysts. Planning required a total of a day and a half of their time, for a total of approximately 36 man-hours. Within this period of time the analysts made their initial contact with the potential client, defining the basic project and system specifications. With this initial data in hand, they then proceeded to break down the tasks necessary to perform the analysis, develop a small network in order to determine the interdependencies of the tasks, schedule the work, determine the man-hours and other resources required, price the tasks, and assign them.

By way of comparison, the amount of planning effort involved to develop an implementation plan for a complex system costing in the neighborhood of $450,000, required approximately 480 man-hours. This involved the planning efforts of three systems analysts for three weeks, and two programmers (for the purpose of defining tasks and estimating required man-hours for programming the new design) for approximately one and a half weeks.

Using the Workbook for Planning

In planning any type of project, the basic functions consist of:

- Defining the objective or product goal (such as done through the use of the System Specification work sheet, previously described);
- Defining and describing the tasks needed to be performed in order to meet objectives;
- Determining the interdependencies and interrelationships of the various tasks, and scheduling them;
- Determining the resources (manpower, materials, equipment, work area, etc.) required to perform the tasks;
- Logically grouping the tasks and assigning them;
- Pricing and budgeting the tasks.

In controlling a project the plans are usually summarized into schedule control charts, budget control charts, and various other formats in order to provide visibility over the three factors that must always be considered by management:

- Schedule Compliance
- Cost Performance
- Technical Performance (such as quality and accuracy)

The previously described Project and System Specification work sheets provide the initial input for the planning effort. In the case of the System Specification, the basic objectives of a new system are defined, thus providing the framework from which tasks can be derived. This system objectives data is also basic material from which the technical performance plan is derived. The brief description of the types of existing data processing and communications equipment, also recorded on the System Specification work sheet, helps in planning for the types of skills that might be assigned to the project. This is true especially if familiarity with certain types of unique equipment is an important factor.

The Project Specification work sheets, by defining the functions to be analyzed, also provide basic input vital to planning project tasks. The "Schedule Factors" portion of the same work sheets provides important input for schedule development purposes, while the information relative to travel and work space also are important input.

In addition to these work sheets, the Workbook's Analysis/Design Plan section also provides additional work sheets useful in the planning process.

The Task Planning Work Sheet

The Task Planning work sheet is shown in Figure 12. In this illustration, an example of how the work sheet would be used in breaking down tasks for a Phase I activity (Systems Analysis and Design) is shown. There are several important considerations involved in breaking down the tasks required to do a job, all of which are illustrated in the example.

The first consideration is that each task should have a separate and unique control number. Once assigned, these numbers will carry through and serve as cross-references for the same tasks in all the other plans, including schedules, prices, and assignments. Any numbering scheme that is convenient and easy for the analyst to use may be applied. However, it should be such a scheme as to allow for the later addition or deletion of sub-tasks without the necessity for renumbering. And the level of any task in the hierarchy should be readily determined from the number. Thus, in the illustration, tasks 1111 through 1115 are all sub-tasks of 1110.

The next important aspect in task planning is that each task title denotes an actual element of work. Each title, therefore, contains an action word, a verb. It is not sufficient to merely state, "Flowchart." An action word has to be added. For example, the terms "Layout Flowchart" or "Reproduce Flowchart" would be appropriate work titles.

The Task Planning work sheet provides for the indenting of sub-tasks under their appropriate main task. This has the effect of making only the lowest level of tasks in each family "tree" actual work tasks. Tasks 1111 through 1115 in Figure 12, for example, constitute the total of all the work required to accomplish task 1110. For the purpose of illustrating this point, if it was such that each of those sub-tasks (1111 through 1115) required one man-hour to accomplish, then task 1110 would require a total of just five man-hours, the sum of its sub-tasks. Another way of expressing this is that each level of work more precisely defines the task it is subordinate to. There is no hard and fast rule as

TASK PLANNING
Analysis & Design Plan

Control No.	Level 0	1	2	3	4	Output
1000	PERFORM SYSTEMS ANALYSIS & DESIGN TASKS					N/A
1100		PERFORM DATA COLLECTION & ANALYSIS TASKS				N/A
1110			DEVELOP SYSTEM REQMTS STATEMENT			N/A
1111				REV. ORIGINAL STATEMENT OF SYST. OBJ'T'S		NONE
1112				DETERMINE INTERFACE REQMTS		INTERFACE WORK SHTS
1113				DET. POLICY & REGULATION REQMTS		P&R WORKSHEETS
1114				DETERMINE USER PREFERENCES		U. PREF. CHECKLIST
1115				DRAFT REQUIREMENTS STATEMENT		REQMTS STATEMENT
1120			DEVELOP EXISTING SYST. FLOWCHART			N/A
1121				EXAM EXIST'G SYST SOFTWARE DOC.		SOFTWARE WORKSHTS
1122				DET EXIST'G SYST ENVIRON FACTORS		ENVIRON WORKSHTS
1123				EXAMIN' PROCEDURES & DESCRIBE SYST		SYST. DESCR. WORKSHTS
1124				L/O EXISTING-SYSTEM FLOWCHART		EXIST'G-SYST WORKSHTS
1130			ANALYZE EXISTING SYST FILES & DOC.			N/A
1131				OBTAIN DOCUMENTS (FORMS, REPORTS, ETC)		DOCUMENTS
1132				DEFINE FILES		FILE DESCR. WORKSHTS
1133				DETERMINE DATA SOURCE & USE		DOC. WORKSHEETS
1134				PEFORM DATA ELEMENT ANALYSIS		DATA ELEM. MATRIX
1140			CROSS-REF DOCUMENTS TO FLOWCHART			CROSS-REF FLOWCHART
1150			DEVELOP COST/BENEFIT BASE LINE			N/A
1151				OBTAIN SYSTEM STATISTICS		STATISTICS
1152				ANALYSE & SUMMARIZE COST/BENEFIT BASLN		C/B BASELINE
1200		OBTAIN INTERIM APPROVALS				NONE (OR "APPROVAL" FORM)
1300		PERFORM NEW-SYSTEM DESIGN TASKS				N/A
1310			DEVELOP NEW-SYSTEM FLOWCHART			N/A
1311				STUDY EXIST'G SYST FLOW, FILES, REPORTS		NONE
1312				STUDY NEW SYSTEM REQMTS		NONE
1313				DRAFT NEW SYSTEM FLOWCHART		NEW SYST FLOWCHART
1320			DEVELOP NEW-DOCUMENT MOCKUPS			DOC. MOCKUPS
1330			ANALYZE NEW SYSTEM COST/BENEFITS			N/A
1331				EVAL NEW SYST CONCEPT AS IT EVOLVES		STATISTICS
1332				COMPARE NEW VS EXISTING		COST COMPARISON RPT
1340			CROSS-REF REQTS, PREF & DOCS TO FLOWCHT			X-REF FLOWCHART
1400		DEVELOP IMPLEMENTATION PLAN				IMPLEMENT'N PLAN
1500		SUBMIT DSN & PLAN TO CUSTOMER				NONE (OR "APPROVAL" FORM)

FIGURE 12: Example of Tasks Broken Down on Task Planning Work Sheet

to the level to which tasks should be broken down. Rather, they should be defined only to that level where the total task can be easily understood and resources (such as man-hours) accurately estimated.

The right-hand column of the Task Planning work sheet helps assist the systems analyst in developing tasks that are output-oriented and thus, from a management standpoint, easily measurable. It is difficult for management to measure the performance of a task defined as "Think About New Design Possibilities" where no output is specified.

In actually applying the principle of developing output-oriented tasks, again it must be kept in mind that outputs must be specified, wherever possible, for only the lowest level of tasks in each task hierarchy. The output column on the work sheet thus merely indicates an "NA" (Not Applicable) for the higher level tasks. It should also be noted on the example that in nearly all cases some type of output has been specified. In a number of instances, this output consists of work sheets from the Workbook.

In some instances it can be useful for the systems analysis planner to utilize a more graphic technique in developing or displaying the tasks required for a project. Such a technique is called a Work or Task Breakdown Structure, and a portion of one such structure is illustrated in Figure 13. The same criteria holds true here as it did for the Task Planning work sheet; each task contains an action word; each task is numbered; each level of tasks constitutes a definition of the task just above it in the structure. An advantage of this graphic form, although cumbersome to construct for large projects, is that task relationships and control numbering schemes can be more easily visualized.

A variation of a Phase I Work Breakdown Structure is illustrated in Figure 14. It should be noted that in this breakdown of tasks the three basic elements of a systems analysis project are defined at the first task level: analyze; design; manage.

Once the tasks have been broken down and defined, task statements can be generated. These are more commonly called "Statements of Work." They describe and define, in narrative form, each of the major task titles appearing on the Task Planning work sheet.

Once the tasks have been satisfactorily defined, the next important planning effort involves determining the interdependencies and sequence of the tasks. A useful technique for this is the network. In the terminology of networks, circles usually represent events (instantaneous occurrences such as "start" or "stop") while the lines represent activities (tasks).

Figure 15 shows a portion of a network which displays a sequence in which the *analysis* tasks, previously outlined on the Task Planning work sheet, can be performed. Figure 16 is a continuation of the same network, in this case showing a possible sequence involving the *design* tasks for the same project.

Once the network has been completed, control schedules (including travel schedules, if applicable) can be developed. Input for the schedule includes the network, and data derived from both the Project Specification work sheet and the Task Planning work sheets. The scheduling task should be done in conjunction with the planning of the resources.

The Resource Estimating Work Sheet

Once the network tasks have been defined and sequenced, the job of estimating manpower and material requirements for the project is the next order of business. To

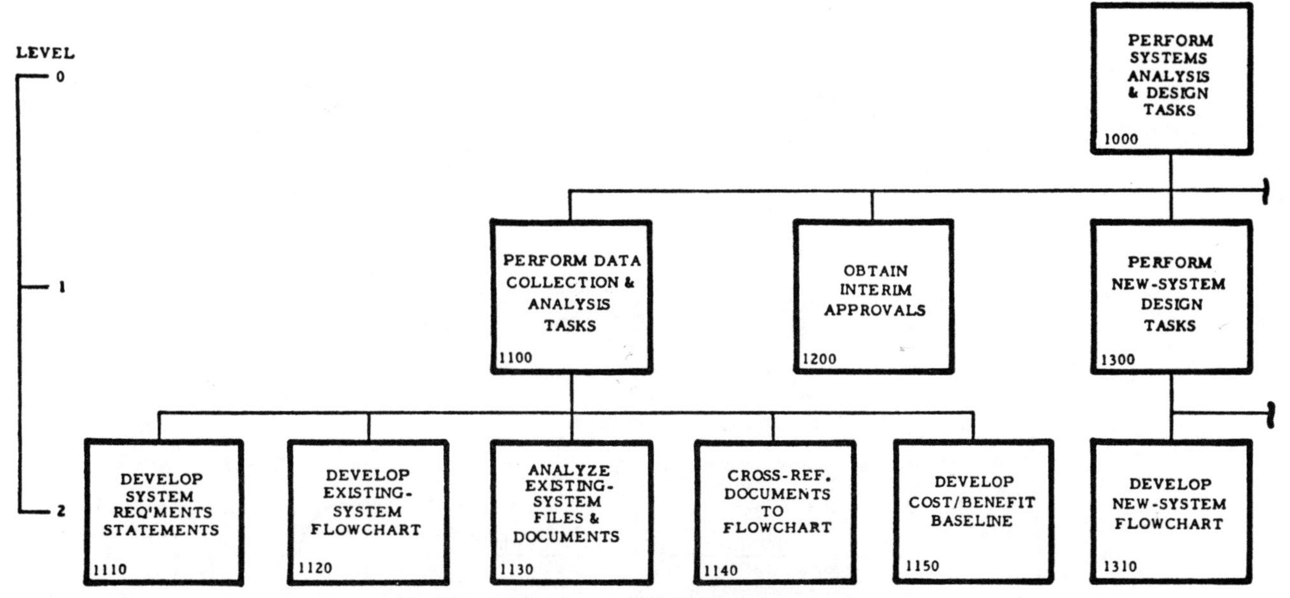

FIGURE 13: Portion of Work Breakdown Structure

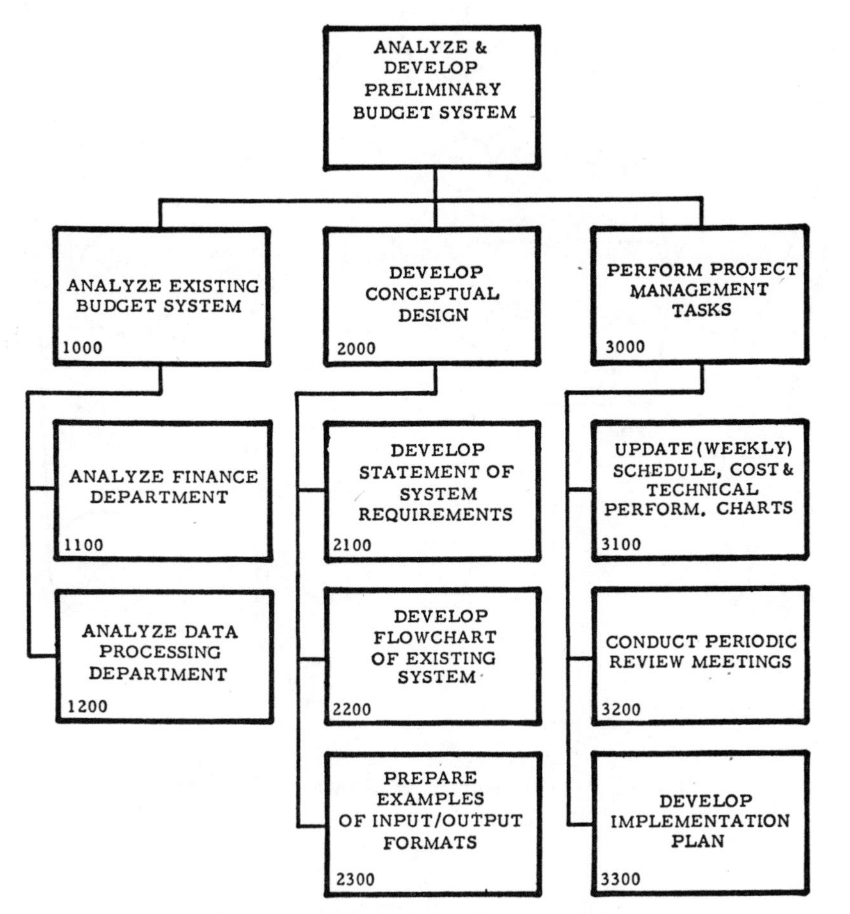

FIGURE 14: Variation of a Phase I Work Breakdown Structure

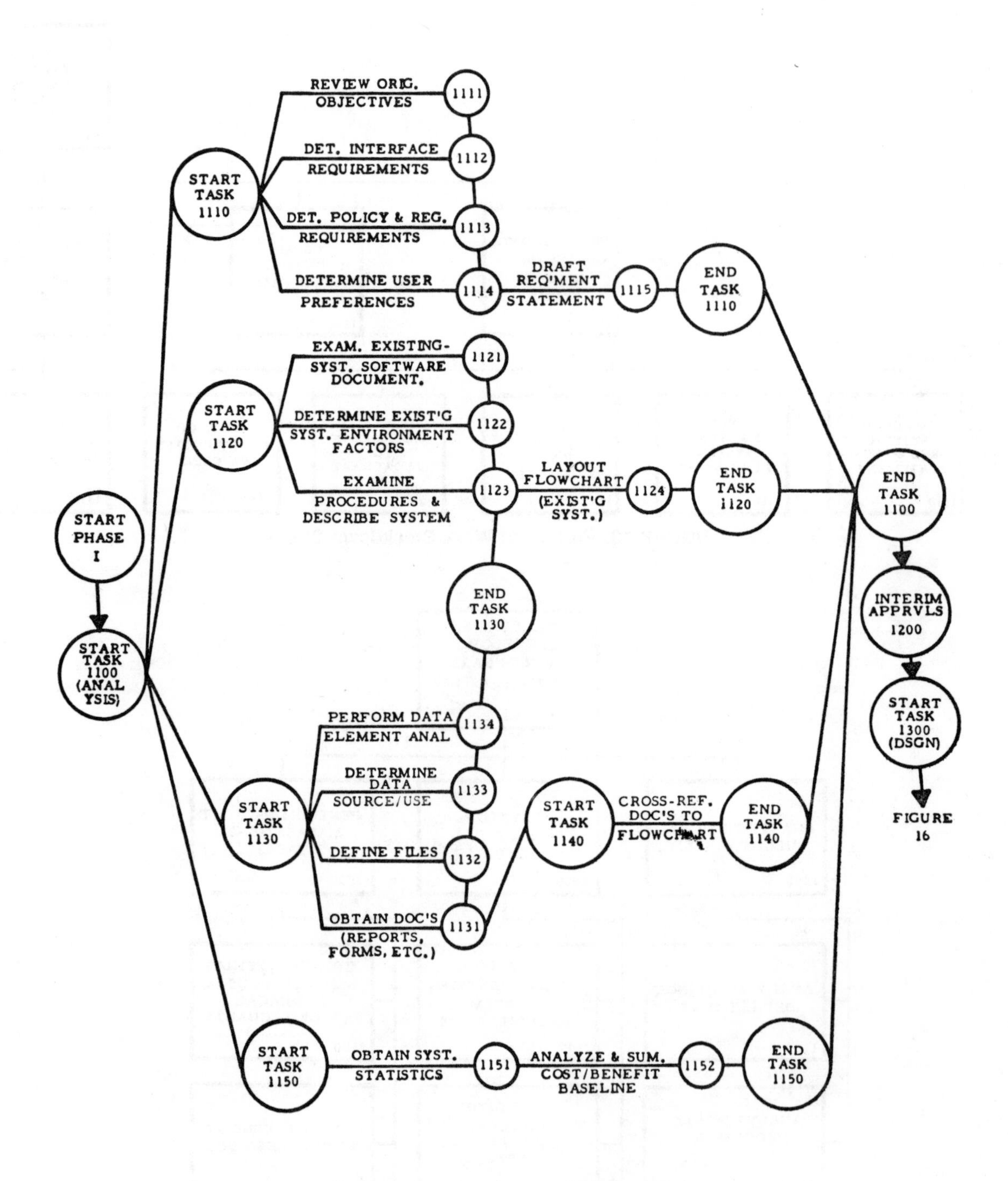

FIGURE 15: Portion of a Network—Analysis Tasks

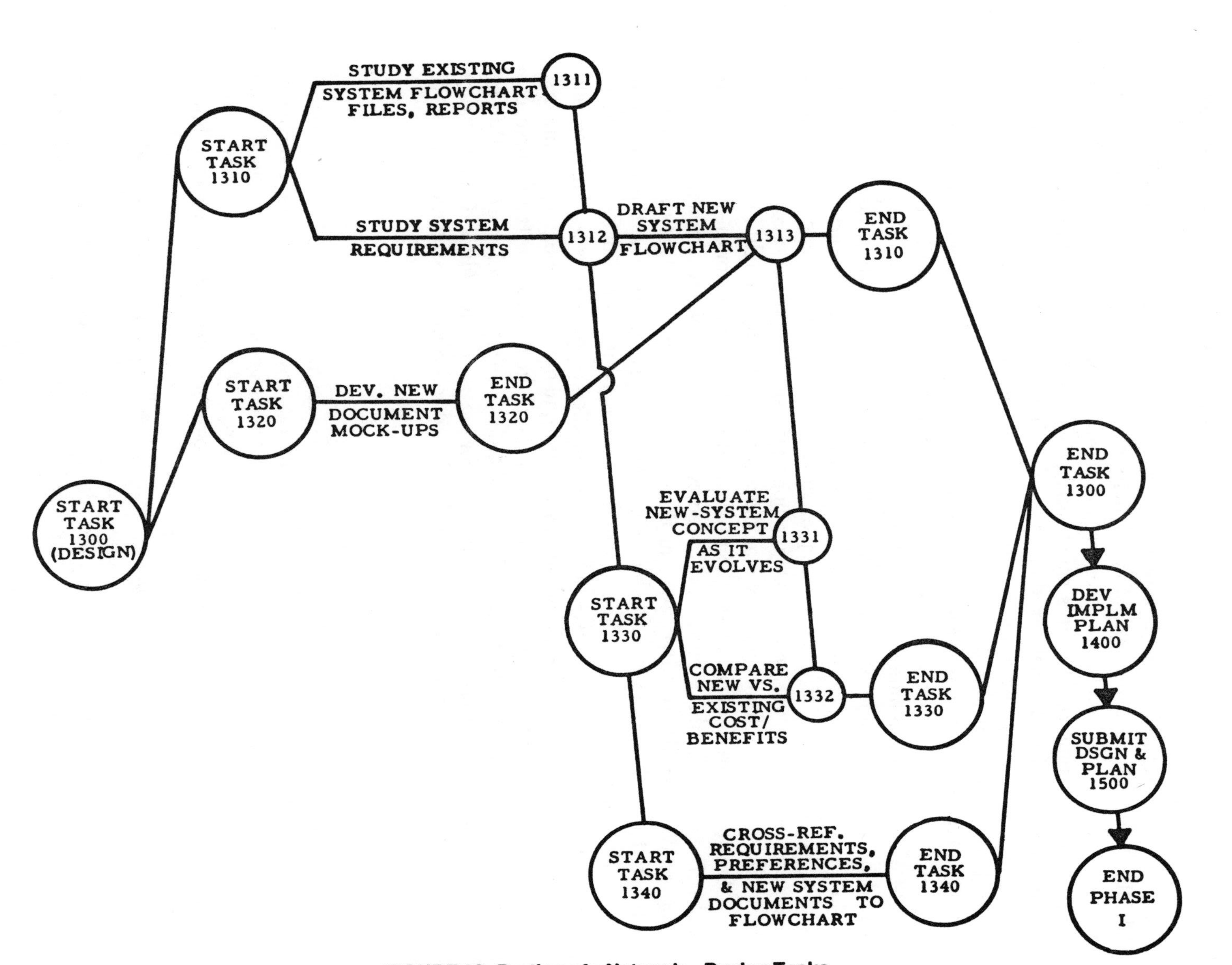

FIGURE 16: Portion of a Network—Design Tasks

assist in this effort the Workbook provides a Resource Estimating work sheet, Figure 17. The illustration shows an example of a work sheet partially filled out with estimating data. In the example, tasks and their corresponding control numbers from the Task Planning work sheet are transcribed to this sheet. Space is provided for estimating man-hours either in terms of individuals or skill type. In either case it must be an itemization according to individual wage rates or categories of wage rates so that costs can subsequently be readily calculated.

The last three columns of the work sheet provide space for noting other material or service requirements that could add to the cost of the project. For a Phase I project this might consist of only a minor amount of reproduction work. In a Phase II project there could be a considerable amount of reproduction required (printing new forms, for instance) as well as a great number of hours required for computer services. Columns should be designated, as in the example, so that different price-range categories of computer services can be itemized. Depending on the specific project, a column designation scheme different from that shown on the work sheet example might be appropriate.

As with the breakdown of tasks themselves, estimating is done for the lowest level of tasks. The reason for this is better illustrated by showing a breakdown of man-hours on a graphic Work Breakdown Structure. In Figure 18 it can be seen that the man-hours estimated at the lowest level in the hierarchy are merely summarized at the level above them. This same principle would be carried out throughout the rest of the structure so that at the top task level there would be one summation of man-hours by skill type, and one summation of all material and services requirements. This one set of numbers, multiplied by their rate factors, would give a total cost of the project. Costs, of course, can also be calculated for a single task, or a group of tasks. Budget control charts can thus be developed according to tasks, groups of tasks, or for the project as a whole. Once tasks are assigned, budget control charts can also be developed according to assignment responsibility.

Technical Performance

As previously described, the system objectives originally outlined on the System Specification work sheet become the prime input in the development of the technical performance plan. Technical performance involves such factors as accuracy, quality, and reliability and can be more easily visualized in terms of the tests and demonstrations called for in a Phase II project. It is more difficult to devise means of measuring technical performance of an Analysis and Design project.

One facet of technical performance control for a Phase I project has already been described. This involves the interim approval, by the client, of certain of the materials produced as a result of analysis. These include the Statement of System Requirements, the existing-system flowchart, the cost/benefit baseline, and the User Preference Checklist.

If the size and scope of the project warrant it, a Design Review Checklist, such as illustrated in Figure 19, can be used. In this example the system objectives would be listed in order of priority in the left-hand column. Next, four review periods, spaced throughout the analysis and design period, are provided for on this particular form. At

RESOURCE ESTIMATING **1.7**

Analysis & Design Plan

Tasks No.	Title (Resources)	Manhours (Indiv. or Skill) a	B	C	D	Repro.	Comp. Serv.	Non Main-frame Comp.
1111	Review Original Objective	4						
1112	Determine Interface Reqmts	40	8					
1113	Det. Policy & Reg. Reqmts	16						
1114	Determine User Preferences	48	8					
1115	Draft Requirement Statement	30			3	10 COPIES 60 PG EA.		

FIGURE 17: Partially Completed Resource Estimating Work Sheet

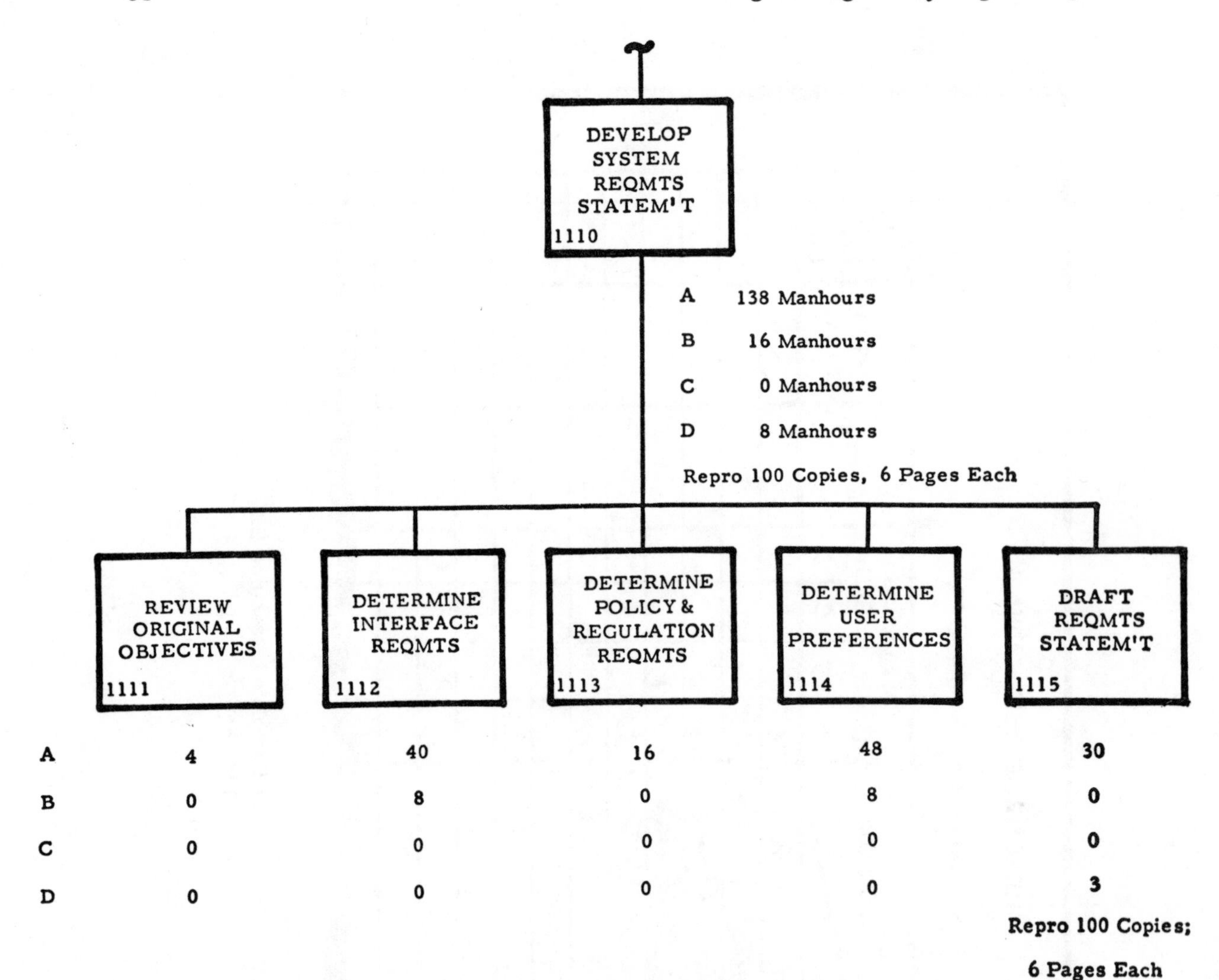

FIGURE 18: Example of Utilizing WBS for Resource Estimating

each review the system designer is asked, in effect, if progress is being made toward achieving each of the system objectives. If not, the required corrective action is noted. Such a checklist, under some circumstances, can serve to keep the goals of the project in focus.

Grouping and Assigning the Tasks

One of the final steps in the planning process is the logical grouping and assignment of the tasks. This is an area of planning, of course, that becomes more complex the larger the project. Large or small, the idea is to match the tasks with the appropriate available skills. These groupings do not need to parallel in any way the Work Breakdown Struc-

DESIGN REVIEW
Checklist

	Review Period	(✔)	Current Status	Action Required
Objective 1:	1			
	2			
	3			
	4			
Objective 2:	1			
	2			
	3			
	4			
Objective 3:	1			
	2			
	3			
	4			
Objective 4:	1			
	2			
	3			
	4			
Objective 5:	1			
	2			
	3			
	4			
Objective 6:	1			
	2			
	3			
	4			

FIGURE 19: Design Review Checklist

1.8

TASK ASSIGNMENT

Analysis & Design Plan

Task No. ____________

Task Title ____________

Task Description ____________

Specific Output(s) Required ____________

The Start Of This Task Depends Upon ____________

Subsequent Tasks Dependent Upon This One's Completion Are: ____________

Schedule:

Start Date ____________

Completion Date ____________

Resources:

Manhours ____________

Materials/Services ____________

Other ____________

Assigned To ____________ Date ____________

Assigned By ____________ Date ____________

FIGURE 20: Task Assignment Sheet

ture. The purpose of the structure is to define the total effort, not to define an organization.

If the project is large or complex in nature, there is justification in using the Task Assignment sheet illustrated in Figure 20. On this sheet all of the basic planning data for a given task is summarized. This includes the task title, the task description, and the specific output expected from this particular assignment. The relationships of this to other project tasks are specified, as well as the start and complete dates for the task, and the maximum amount of resources that the assignee is expected to utilize in accomplishing his work.

Submittal and Approval

In summary, once the planning process has been completed, the systems analyst is in a position to submit for client approval the following data:

- A Statement of System Objectives
- A Statement of Work (List of Tasks)
- A Network/Schedule
- A Resource Plan (including project cost)
- A List of Assignments

In addition, he is in a position to submit the various aids that can be used in controlling the project, including schedule control charts, examples of cost or budget control charts, and technical performance control charts or lists.

3
Establishing Successful Relationships and Determining the System Environment

The way the systems analyst handles himself during the first several days of the project can be of vital importance to the project's ultimate success. The key problem to overcome involves communications: communicating a clear understanding of the project's purpose and objectives, its scope, and the methods to be employed in conducting the study.

By the time the project is ready to start, there is already such an understanding with one or several members of client management. Objectives and other elements of the plan as to how the analysis and design effort will take place have been formulated and agreed to. Unfortunately, though, in many cases this has not yet been communicated to other members of management or to all the client's employees who will be affected by the analysis.

People whose work is to be the subject of an analysis are sometimes confused, usually apprehensive. Up until this point they have probably heard only rumors and, possibly, distorted ideas of what the subject and purpose of the analysis is. There is the case of the chaplain at a Navy base whose function, along with all the other base activities, was the subject of an analysis for the purpose of developing an improved reporting system. His confusion regarding the project was evident by his opening remark to the systems analyst, "What are you studying my function for? How can you automate God?"

It is normal to expect some supervisory personnel to be worried that their skills and personal work techniques will be subjected to criticism. Other employees, also not understanding that the main thrust of the study is an analysis of the system rather than the individual, are fearful that their jobs might be in jeopardy. These fears, if unchecked, can seriously undermine the project. The client is usually concerned, too, that the analysis proceeds productively, but without unduly upsetting the employees.

The size of the organization to be studied usually dictates the communications approach the systems analyst can use. In many cases, a first-day briefing permits the analyst to meet with a number of the client's management and supervisory personnel. This can be in the form of an indoctrination briefing where the client personnel briefly describe their organization units, functions and responsibilities, and the systems analyst reviews the system objectives, the project plan, and the analysis approach that is to be used.

Establishing good working relationships with the rest of the employees usually must depend on personal interviews, a subject more fully covered in the next chapter. It is from these first interviews, and from a combination of other sources (including the indoctrination briefing) that the analyst picks up the preliminary data regarding the environment in which the existing system operates (organization structure, pertinent history, forecasts, and other such information).

INDOCTRINATION BRIEFING

It is quite common for indoctrination briefings, sometimes called familiarization briefings, to be conducted on the first day of a new systems analysis project. These briefings are normally requested by the systems analyst well in advance to permit all concerned to prepare for it. The briefing affords the opportunity for the analyst, or analysts (as the case may be), to be personally introduced to key personnel in the client organization. It affords the opportunity for an exchange of pertinent information, and provides a setting for informal discussions which help clarify approaches and goals.

Of prime interest to the systems analyst, because this is a source of pertinent system environment information, are the client-personnel descriptions of the missions of their organization units, organization structures, history, and descriptions of the various products or services offered. Of importance, too, is a general description of the system or sub-system which is the subject of the analysis and design effort.

INTRODUCING THE WORKBOOK

The indoctrination briefing, as previously mentioned, is really a two-way street, for it offers the systems analyst the opportunity to communicate with key client personnel. The analyst can explain the previously agreed-to objectives of the analysis and design effort, and can summarize key points from the plan, including the functions which are to be studied. If there is a team of systems analysts, they can be introduced, and an explanation provided as to areas of assigned responsibility.

Also a possibility at this time is an explanation of the use of the Workbook. The systems analyst can explain how the Workbook is used, and show how the workbook approach is one that concentrates on the system rather than on the individual. However, too much time explaining the Workbook might also serve to intimidate the prospective interviewees by making the process seem too mechanical and doctrinaire. The amount of emphasis to be placed on this subject will depend on how the analyst sizes up the audience and how well he feels they will understand the overall process.

It is also essential at this time for the systems analyst to explain that as various work sheets are filled out, and as other materials (such as flow-charts) are generated from this information, the people who provided the original input will be allowed to review the information before it is finalized. This is to assure the data's correctness, completeness, and accuracy. Furthermore, if the generated material is useful to the employees' current work, as sometimes is the case, it can probably be arranged that they receive working copies of that material. These approaches in the use of the Workbook, once understood by the employees, usually serve to set their minds at ease.

If the project is large in scope and there are a great number of employees to be interviewed, it is often useful to develop an appointment schedule. If the key members of the organization are in attendance at the first-day briefing, this often is an opportune time for making such arrangements. Once arranged, formal notification of the schedule can be distributed to the affected personnel.

DOCUMENTING SYSTEM ENVIRONMENT FACTORS

The second main section of the Systems Analysis Workbook contains work sheets useful for recording data relative to the environment in which the system operates. This includes information relative to various aspects of the organization and its structure, trends and forecasts, pertinent historical data, and a detailed description of existing and planned data processing and communications equipment.

The intent, of course, is to describe the environment in which the system operates. This is information over and above that which describes the system itself. It's the type of data that provides the analyst with the basic orientation required for subsequent steps in the study. It also provides him with information useful in anticipating future system capacity requirements.

It is possible that, prior to the interviews, some or all of this information may already have been recorded on the work sheets. Some of this information could have been derived from the initial contact with the client, subsequent research, or from material presented at the indoctrination briefing. The job here, then, is to make certain the information is complete, filling in any missing elements.

Organization Work Sheet

A series of work sheets is provided in the System Environment Factors section of the Workbook for the purpose of recording information relative to the organization. The first two of these are the Organization Identification work sheet and the Sub-Unit Identification work sheet, both similar in format.

In the first instance, the Organization Identification work sheet, Figure 21, provides space for identifying the study's prime organization by all the various names and codes under which it might be known, and for recording a brief description of the organization's

ORGANIZATION IDENTIFICATION

Formal Organization Name ____________________

____

Organization Number or Code ____________________

Standard Abbreviation or Acronym ____________________

Other Names Used ____________________

____

Names of Preceding Organizations ____________________

____

This Organization is directly subordinate to (Organization Unit) ____________________

____

____

Capsule Desription of Organization Prime Mission (Attach Organization Chart. Chart should contain names and titles of key personnel) ____________________

FIGURE 21: Organization Identification Work Sheet

prime mission. Identification is in terms of the organization's formal name, any organizational number or code, any standard abbreviation or acronym which is used to identify the organization, any other names used, or any former names used. If the study's prime organization unit is other than the company as a whole, the name of the organization to which this unit is directly subordinate is also called for. This type of identifying information is important to the analyst. Later, when reviewing procedures, policies, regulations, and other similar material, if the organization is cited through the use of something other than its formal name, the systems analyst will still be able to correctly identify it.

Space is also provided on the work sheet for briefly describing the organization's prime mission. This would be a capsule description of the organization's primary functions and responsibilities. If an organization chart can be obtained, or if information can be gathered from which the systems analyst can sketch such a chart, that chart should be appended to this particular work sheet. As noted on the work sheet itself, the organization chart should contain names and titles of key personnel.

The Sub-Unit Identification work sheet, Figure 22, provides space for recording the same categories of information as just specified for the prime organization. Separate work sheets, as noted, should be prepared for each sub-unit that falls under the scope of the system being studied. Again, it is of importance to either obtain or develop an organization chart and append it to each of these Sub-Unit Identification work sheets.

A work sheet is also provided for listing key personnel in the client organization. This work sheet is shown in Figure 23. This is the type of information that aids the systems analyst in performing subsequent steps in the analysis and design process. He should list, on this form, those key people in the organization whose functions and responsibilities are involved with the system under study. The form provides space for identifying the person's name, telephone number or extension, organization unit, title or position, specialty, and location.

The "specialty" category on this work sheet is of particular interest. Every systems analyst, as he conducts interviews and collects data, picks up valuable leads as to sources of vital system information. He hears remarks such as, "You should talk to 'so-and-so'. He knows all about that problem." It is on this Key Personnel work sheet that the analyst records these leads, noting the "specialty." In one actual instance involving a systems analysis of a large Wall Street institution, through leads such as just described, a man was located in the organization who had been with the firm 35 years. It turned out he knew and understood every single aspect of the existing system, how and why it had been instituted, and could describe detailed problems with the system that were beyond what others suspected. He could describe two generations of previous systems, and was able to construct an articulate genealogy of nearly every form and report used in the system.

The Headcount work sheet shown in Figure 24, provides the systems analyst with one measure of system growth and growth requirements. This type of information aids the analyst in his design effort in terms of planning a system that will accommodate possible growth in volume and capacity. The work sheet provides space for recording, by sub-unit, headcount figures for the past, present, and future. The systems analyst can select appropriate time periods for the past and future so that a trend can be seen. Space is also provided for summing up the total headcount for the organization under study, and for noting any comments appropriate to the subject of headcount.

SUB-UNIT IDENTIFICATION

(Prepare Separate Work Sheet for Each Unit)

Sub-Unit Formal Name________________________________

Sub-Unit Number or Code________________________________

Standard Abbreviation or Acronym__________________________

Other Names Used______________________________________

Names of Preceding Organizations__________________________

This Organization is directly subordinate to (Organization Unit)__________

Capsule Description of Unit's Prime Mission (Attach Organization Chart)_______

FIGURE 22: Sub-Unit Identification Work Sheet

KEY PERSONNEL

Name ______________________________ Phone ______________
Organization Unit ______________________________
Title or Position ______________________________
Specialty ______________________________
____
Location ______________________________

Name ______________________________ Phone ______________
Organization Unit ______________________________
Title or Position ______________________________
Specialty ______________________________
____
Location ______________________________

Name ______________________________ Phone ______________
Organization Unit ______________________________
Title or Position ______________________________
Specialty ______________________________
____
Location ______________________________

Name ______________________________ Phone ______________
Organization Unit ______________________________
Title or Position ______________________________
Specialty ______________________________
____
Location ______________________________

Name ______________________________ Phone ______________
Organization Unit ______________________________
Title or Position ______________________________
Specialty ______________________________
____
Location ______________________________

FIGURE 23: Key Personnel Work Sheet

	Past (19)	Present	Forecast (19)
Unit ______	______	______	______
Unit ______	______	______	______
Unit ______	______	______	______
Unit ______	______	______	______
Unit ______	______	______	______
Unit ______	______	______	______
Unit ______	______	______	______
Unit ______	______	______	______
Unit ______	______	______	______
Unit ______	______	______	______
Unit ______	______	______	______
Unit ______	______	______	______
Total Organization	______	______	______

Comments:

FIGURE 24: Headcount Work Sheet

Describe plans or trends in this industry or field that could affect this system ________

Describe plans or trends in this organization that could affect this system

Changes in functions or responsiblities ________

Change in location ________

Unique location characteristics ________

Changes in size or workload ________

Other ________

FIGURE 25: Plans and Trends Work Sheet

Forecast factors, such as headcount, are of significant importance to the system designer. Another work sheet which will assist him in his effort to design a system with appropriate consideration for anticipated future developments, is the Plans and Trends work sheet shown in Figure 25. It calls for information relative to anticipated changes in the industry or field, and in the organization's mission, size, or location. A plan to move to a new facility, for instance, could significantly impact a new system concept. In describing location information, there is also space for noting any unique location characteristics either existing or planned. Is the data processing equipment used in the system located across town, for instance?

If official business forecasts involving the organization being studied exist, they should be appended to this work sheet, or placed in the appendix section of the Workbook and referenced on the work sheet.

Historical Data Work Sheet

Background information relative to any significant historical activity that may have affected the existing system can be recorded on the Historical Data work sheet shown in Figure 26. In this case, similar to Plans and Trends, information is recorded relative to any major changes in the organization's mission, size, or location. Now, however, it is in terms of the past, especially the recent past. Reasons for such changes should be noted, if such information is available, and problems encountered should be described.

Equipment Description Work Sheets

Equipment Description work sheets are provided in the Workbook. The original System Specification work sheet briefly described existing data processing and data communications equipment for the purpose of developing the initial project plan. The Equipment Description work sheets call for the describing of this hardware in much greater detail. Figure 27, for instance, displays a work sheet useful in collecting data relative to the existing data processing equipment. Information such as make and model, number of units, storage types and capacity, and usage is called for. The point, here, is not to secure an equipment specification (which can be secured from the manufacturer), but rather a description of the equipment as it is now being used, especially in relation to the system under study.

The work sheet illustrated in Figure 28 provides space for recording data relative to other types of equipment (data communications equipment, reproduction equipment, business machines). As noted on the work sheet itself, this should be used only in those instances where this type of equipment is significant to the operation of the existing system.

Finally, information as to any types of data processing equipment that may be on order is to be recorded on the work sheet illustrated in Figure 29.

Describe past changes in this industry or field that did or could affect this system

Describe past changes in this organization that did or could affect this system

Changes in functions or responsibilities

Change in location

Unique location characteristics

Change in size or workload

Other

FIGURE 26: Historical Data Work Sheet

EQUIPMENT DESCRIPTION

Existing Data Processing Hardware

Make & Model ________________________________ No. of Units ________

Location ____

____

Owned? Leased? Time Shared? Service Center? ____________________

Storage Type and Capacity __________________________________

____

____

Operating Features ____

____

____

____

Assemblers/Compilers ______________________________________

____

Service Routines ____

____

Usage for System Under Study ________________________________

____

____

________________________________ Estimated Hours or % Usage ________

Input Equipment:

Type	Make & Model	No. Units

Output Equipment:

Type	Make & Model	No. Units

Describe Other Peripheral/Auxiliary Equipment ____________________

____

____

FIGURE 27: Equipment Description—Existing Data Processing Hardware

EQUIPMENT DESCRIPTION

Data Communication, Reproduction, Business Machines

(List only those items significant to operation of existing system)

Other Data Communications Equipment (TWX, Facsimile, Etc.)

Type	Make & Model	No. Units	Hrs. or % Use

Describe Usage As It Applies To System Under Study

____

____

Reproduction Equipment (Duplicating, Printing, Etc.)

Type	Make & Model	No. Units	Hrs or % Use

Describe Usage As It Applies To System Under Study

____

____

Business Machines (Calculators, Addressing Equipment, Etc.)

Type	Make & Model	No. Units	Hrs or % Use

Describe Usage As It Applies To System Under Study

____

____

FIGURE 28: Equipment Description—Data Communications, Reproduction, Business Machines

EQUIPMENT DESCRIPTION

Data Processing Hardware On Order

Type of Equipment on Order (Make, Model No., Quantity) ____________

Prime Purpose of New Equipment ____________

When Ordered ____________

Delivery Due ____________

Installation Considerations ____________

Disposition of Existing Equipment ____________

Other Considerations ____________

FIGURE 29: Equipment Description Work Sheet—Data Processing Hardware on Order

4
Planning and Conducting Result-Getting Interviews

The most important ingredient in any system analysis is the body of fact on which it is based. This body of fact must be complete; it must fully describe the system which is already in existence and the environment in which it operates. Although an essential part of it comprises the forms and documents being used, these alone are not sufficient. The ultimate source of the critical facts is the people who are part of the system, the operators, the users, those who input the information, and the system managers. The only efficient way to obtain the required information is to ask these people; that is, to conduct a series of interviews.

In planning, conducting, and following up an interview, the Workbook can be an invaluable aid. Because it is designed to organize all of the information needed for a successful analysis, it serves also to organize the data collection tasks. By reviewing the Workbook before the interview, the interviewer will be reviewing the elements of information he needs and also the logic and reasons behind their necessity. If he has the Workbook in front of him during the interview he will not only be able to check off and record the information as the interview progresses, he will have a tool for redirecting the interview if it begins to stray. Finally, the blanks remaining in the Workbook after he has recorded the results of his interview will point up the data that is missing and enable him to plan follow-up effort.

Collecting the various categories of system data is the subject of other chapters in this book. This chapter concerns itself with the essential elements of interviewing practice, including planning interviews, conducting interviews, and principles of successful interviewing.

THE INTERVIEW PLANNING CHECKLIST

An interview must be planned. The systems analyst must know why he is there and what information he is after. Prior planning will not only insure that the needed information is obtained, it will also enhance the respect paid the interviewer by the interviewee; it is the professional approach.

As in any planning, the first step is to determine the objective of the interview. This will depend not only on the analyst's requirements and the project objectives, but on the person being interviewed. He may be principally a user of information. Or he may be significant as a source of information to be input to the system, or he may be an operator or administrator. A review of the Key Personnel work sheet, particularly the notes entered in the "Specialty" block, should be made while considering the interview objectives. This will tell the analyst what unique information the respondent might have, as well as what his function is in the organization.

To aid in the other part of planning an interview, that is, in determining what information is needed, it is well to review the tasks that are ahead during the rest of the analysis phase of the project. These tasks determine the requirements that the analyst will have for particular information.

INTERVIEW PLANNING **1.10**

Checklist

Interviewee (Name)________________

Organization Unit________________

Title or Position________________

Interview Date/Time________ Follow-Up________ Review________

(Indicate, by check mark, topics essential to this interview)

Topic	Work Sheet	Check	
		Interview	Follow-Up
Organization Identification	2.1		
Sub-Unit Identification	2.2		
Key Personnel	2.3		
Headcount	2.4		
Plans And Trends (Forecasts)	2.5		
Historical Data	2.6		
Equipt. Description-Exist'g Data Processing	2.7		
Equipt. Description-Bus. Machines, etc.	2.8		
Equipt. Description-Hardware On Order	2.9		
Interfaces--Inter-Organizational	3.1		
[illegible] Report	3.2		

FIGURE 30: Portion of Interview Planning Checklist

To assist in this aspect of planning, the Interview Planning Checklist has been included in the Workbook as the last form in the Analysis/Design Plan section. This checklist, shown in Figure 30, consists of a list of topics which must be covered if all the information needed is to be obtained. The topics on the checklist are cast at a more general level than those which the analyst will be asking during the interview. They are not intended to be used as an interview script but to indicate the essential categories of information which he should be pursuing. The topics are organized according to the major sections of the Workbook and, so sequenced, follow the general order in which the data most likely will be collected.

In using the checklist, and prior to the interview, the systems analyst should review the data on the Key Personnel work sheet and whatever else is known about the person to be interviewed. At the same time he should scan the topics on the list and check those for which he feels the respondent would have significant answers. These will indicate the areas on which emphasis should be placed during the interview. Included with the topics on the checklist are the numbers of the Workbook pages on which each type of data is recorded. It would be well for the analyst to flag the pages associated with his key topics so that he will be able to turn to them quickly during the interview itself.

CONDUCTING THE INTERVIEW

The systems analyst should condition himself to think in terms of the essential elements of information. Once in the interview situation, these should be kept in mind; referring to them mentally keeps the interview on the right track and ensures against missing some critical piece of information. But they should not be used to create an inviolable script. The interviewer must be opportunistic in pursuing information. Let one answer determine and shape the next question.

Explaining the Purpose and Approach

As the first step in the interview process, the analyst should explain his purpose. Although this may seem obvious, it can be critical. It sets the stage for the rest of the interview. It is essential to lay the foundation for the confidence and respect that is so important to success, and this can be done right at the outset.

The analyst should emphasize that he is there to get facts regarding the current system's operation. The idea is to improve that system, or devise a new one, so it will be more useful to the people who must use it. It is important that the respondent hear this from the analyst even though he has probably already been briefed.

In sizing up the interviewee, the analyst may decide, at this point, that it could be beneficial to show the interviewee the Workbook and explain its function and contents. It can also be explained that information developed by the analyst will be checked back with the interviewee. Thus the analyst will be able to insure that the information being used in the analysis will be factual.

This will also serve to build confidence and trust on the part of the respondent. It will assure him that the interview is not for the purpose of evaluating him as an individual, nor will it produce a secret, critical report for his employer. By promising him a review of the recorded material, the analyst goes a long way to establishing the cooperative relationship that is essential to a successful interview.

Giving the respondent a chance to review the Workbook data also involves him more seriously in the project and fosters a pride in authorship that may pay real dividends to the analyst later.

Questions and Answers

The first question after the purpose and approach have been explained can also be an awkward step if not thought out ahead of time. This is the time of beginning, of getting down to business. It is a transition to the real work and therefore should be made as easy as possible. A good place to start is with the respondent's position in the organization and his responsibilities or mission. This gives him an easy question to answer and at the same time allows him to express his own personality.

Moving from his organizational position to his responsibilities, the mission of his organization or the tasks he performs is an easy and logical step. It is logical, then to move on to the information he needs to perform each task, the way he gets this information, and the systems that support his efforts, that is, the flow.

From here on the interview will generally take its own course. The analyst will be guided by the pages he has flagged in the Workbook and by following the flow of information through the system. The respondent himself will also have some control over the direction the interview takes. The analyst should follow the respondent's interests as much as possible so long as these interests don't take the discussion off on too much of a tangent. When his interest flags or it is necessary to get back on course, reference to the Workbook will generally be sufficient to keep the interview going.

The analyst should be ready to ask such questions as: "How does this work out in practice?" "Does that report really give you what you want?" "Do you think this could be done better or differently?"

Respondents will differ in both their willingness and their capability to give information. Some will be hard to stop once they get started and questioning will be at a minimum. Others will have to be led, step by step, through the interview with each question eliciting a single, bare response. The words "how" and "why" are invaluable aids in situations like these.

For the talkative respondent who goes off on a tangent or becomes too interested in a pet subject, it is a poor tactic to cut him off abruptly. One approach is for the analyst to suggest that another visit be scheduled in the future devoted to this subject only.

Any sensitive information the client has deemed unnecessary to the study has been previously defined as a project constraint or limitation. All other information essential to the analysis should be made available to the analyst, and this fact explained to the interviewee. It should also be explained that all sensitive information will be handled with discretion.

Throughout the interview there will be references to various forms, reports, and standard documents. Obtaining permanent copies of the forms and reports is essential. Other documents (such as sets of procedure books, for example) are more difficult to obtain on a permanent basis. Arrangements must be made, then, to borrow these.

Taking Notes

Taking good notes is very important; it is also awkward. People react in different ways to having their words taken down. Some freeze; others start dictating. Using the Workbook to record data during an interview alleviates much of this awkwardness and also helps in keeping the interview on the track. It can be useful, too, to use a recording device if this can be done in an unobtrusive fashion so as not to inhibit the interviewee.

If the analyst has flagged the pages which he expects to be most pertinent, it will be easy for him to follow his interview plan. Being able to look at the entire page rather than just a notation of a subject area also helps in making sure all the needed data about that area is obtained.

Actually, the notes taken during an interview need not exactly fit the worksheets. The important thing is to get the facts recorded while they are still fresh. It helps if the information is recorded as close as possible to where it should ultimately be and in the right format. However this is not always possible. For this reason it is wise to make pencil notes in the margin which can be transcribed in the right form later, or to use a rough work sheet and transcribe to a clean copy.

TRANSCRIBING AND PLANNING FOLLOW-UP

The interview notes should be entered in final format in the Workbook as soon after the interview as possible. It is surprising how much that is heard and understood during the interview becomes a mystery or a blank the next day. The notes must of necessity be taken in some sort of abbreviated form. Very often such abbreviations, so obvious when set down, become mysterious signs later.

It is particularly important to complete the Workbook pages quickly if more than one analyst is using the master Workbook. Everyone on the team should have access to new information as soon as it is available.

Finally, immediate transcription allows the analyst to plan and schedule his follow-up interview as early as possible. After the notes are transcribed, the analyst will be able to see any blanks which should be filled in. It is the presence of these blanks, representing gaps in the needed information, that will influence the planning of the follow-up interview. The analyst should also review the Interview Planning Checklist, marking those questions that have been answered. Check marks can then be made in the follow-up column to indicate new topics to be covered, or old topics in need of review.

A follow-up appointment will be necessary, in any case, to review the data with the respondent. For this purpose the work sheets should be as neat, accurate, and complete as possible.

PRINCIPLES OF SUCCESSFUL INTERVIEWING

The interviews are critical in the data collection process; they are also critical in that the essential relationships with the organization being studied are established by them. An arrogant or flippant approach to the questioning can kill a project no matter how fine the design of a new system. Even if a system is installed it will not succeed in the face of antagonistic users, or even more to the point, of the antagonism or resistance of those who must provide input. During the interviews an image of sincerely wanting to solve the problems of the people being interviewed should be established. This will enhance the success of any system ultimately installed.

The Importance of Listening

It is essential to remember that the purpose of any interview is to get facts. This in turn requires listening. An analyst is not receiving any information so long as he is talking himself. Granted he must ask questions; he need not do much talking otherwise. One measure of the efficiency of an interviewer might be the amount of time he spends listening in proportion to the amount of time he spends talking.

"Listen to your respondent" is perhaps the most important yet often violated rule of interviewing. The interviewer should listen even if the respondent is explaining something that is already known. His description may contain subtle overtones that are as significant as the facts themselves.

There is nothing that cuts an interview faster than to be impatient and stop a respondent in the middle of a sentence. The interviewer should not put words in his respondent's mouth, nor interrupt with a summary of what he has said. If the question was important enough to ask, the answer should be treated with the same importance.

A One-to-One Relationship

Success of the interview depends on establishing the right kind of personal relationship with the respondent. This is easiest done when the analyst interviews alone. An interview should approach friendly conversation. The accommodations that must be made by interviewer and interviewee to achieve this are much more easily made when there is no third party to worry about.

An interview also follows a particular logic path. It should be the personal logic of the interviewer. Very often it is important to build slowly to an important question. A second interviewer invariably interjects a tangential question, based on his own logic, diverting the questioning away from the point, often permanently.

There is another reason for the one-to-one relationship, also. The easiest way to over awe a respondent is to descend on him with a whole team of interviewers. It is hard enough to establish the level of confidence needed without overpowering the respondent from the moment the analyst enters his office. Even when the interviewee has a whole team on his side, as well, this creates a difficult situation. He will most likely concentrate on impressing his peers; he knows that they will be evaluating him.

Fact-Finding

The job is to obtain facts. It is important for the systems analyst to avoid voicing criticisms, or spot-evaluations, of what he sees or hears. There is no surer way to kill an entire project than to get the principals worrying about such matters. Nor is there any surer way for the systems analyst to insure that he will get only half the facts he needs.

On the other hand the analyst may encourage his respondent to evaluate and criticize. His dissatisfactions are important clues to real possibilities for improving the system. His criticisms have a validity because he is actually involved in operation or use of the system. They need not be taken at face value but they should be noted and investigated.

Interviewees may use non-quantified terms like "often," and "seldom," and phrases like "very good," and "very bad." These are usually coarse and unfiltered judgments which sometimes don't hold up under objective scrutiny. The analyst should try to acquire independent data and check on the validity of the judgments when they pertain to really substantive matters. He should ask for examples, statistics, and sample data, and he should also check with other sources in the company.

It should also be kept in mind that, very often, the person who is involved in some detail of the operation and who can provide valid information regarding this particular detail, has a poor and probably non-objective viewpoint of the over-all system. Further, interviewees will sometimes be in error about what people several levels down in the organization are doing relative to the system. The analyst should try to double-check critical data selectively.

Being Honest About Lack of Knowledge

Beginning at the introduction and continuing throughout the interview, the analyst should be honest about his lack of knowledge of the interviewee's field of endeavor. If the analyst is not an expert in this field, or has never operated in this type of organization before, he should let the respondent know. The analyst should not try to bluff.

If it so happens the analyst does, indeed, have specialized knowledge in this field, it still is best to play this down. If not, the respondent may start assuming the analyst knows more than he really does and therefore skip over basic facts that are needed.

A lesson in being honest can be derived from the experience of a systems analyst at a major naval base. A key respondent was the operations officer of a fleet unit. He was a career officer who knew his business backwards and forwards but he was stiff and ill at ease in the interview situation. Nonetheless, he talked rapidly and in detail about the flow of command and administrative reports, messages, and directives. Unconsciously he talked a navy language and his descriptions were liberally sprinkled with references to cincpacfleet, subpac, crudespac, and subrons. He spoke glibly of chopping to wespac, of availabilities and ORIs, of shipalts and marlexes.

The analyst did the best he could but that wasn't good enough. After the interview he was able to construct a crude flowchart, but much was to be desired. A second interview was arranged. The officer was just as stiff as ever as he asked, "Well now, what needs clearing up? Just where would you like to start?"

The answer was right to the point. "At the beginning." The analyst explained that he wasn't a sailor and that, frankly, the acronyms and jargon confused him.

It was the right answer. The commander laughed and relaxed. He was careful to explain all of his terms and apparently enjoyed the second interview immensely. The flowchart was fifty percent wrong when they started but was flawless when the interview was over.

Had the analyst tried to bluff his way through in order to impress the commander with his nonexistent knowledge, the entire project might have been jeopardized. On the other hand by confessing his lack of knowledge, he gained the respect and confidence of the respondent and, in fact, overcame the commander's own insecurity about the interview.

Observing Management Channels

Finally the interviewer must respect the organization and policies of the group he is working for. He will normally be shown every courtesy, at least by those who retained or employed him. He is not in a position to demand more. Although the client is dependent on him for solutions, the systems analyst is dependent on the client's people for facts. If he has a legitimate need for more information than he has been given, he should make a proper request, using formal management channels if necessary, and be prepared to explain his reasons for the request.

Observing management channels is a delicate matter sometimes, but an essential one. The analyst is dependent on upper level management for either his contract or employment, whichever the case may be, and dependent on all levels for his facts. Neither operator nor manager should be given reason for hostility toward the analyst if it can be avoided. The analyst may have to be firm but he should use discretion and diplomacy. It is usually wise to handle extraordinary requests or conflict at the lowest level that has proper authority. This avoids embarrassing the worker or lower level manager. The analyst should never threaten to go to a man's superior. At the extreme, he may suggest that this might be a way to resolve a difference, or inform him that he plans to seek a ruling from someone else. If possible he should invite the man to accompany him. Occasionally it will help if the analyst points out that this will relieve the reluctant respondent of responsibility for any ill consequences of granting the request.

These fundamentals are only those of good business relations in any situation. Observing them will ensure and maintain sound relationships during the data collection phase and set an essential pattern for later steps of the project.

5

Determining Necessary System Requirements and Desirable Features

If any single aspect of systems analysis can be singled out as more important than others, it is that part dealing with determining the requirements with which a system must comply. At the beginning of a project, a preliminary step in this direction is taken when a statement of system objectives is formulated. These original objectives, though, are rarely adequate or complete enough to guide the design of a new or improved system. It is necessary, then, to take a penetrating look at controlling regulations, policies, and information-exchange needs. From this detailed study the original objectives can be expanded into more specific and precise terms.

As the analysis progresses, new-system features and capabilities will be derived. Once determined, these can be divided into two categories: those that are necessary and those that are desirable. Those that are necessary and which *must* be incorporated into the new or improved system design are itemized on a Statement of System Requirements. Those that are desirable are listed as "User Preferences," and are evaluated on the basis of costs and benefits. If they pass muster, and if the client-management desires these features, they can also be incorporated into the design.

Requirements are often interrelated. For the purpose of defining and recording them on work sheets, however, it is useful to break them down into types even though there is a risk of some amount of duplication. The Workbook provides work sheets and checklists relative to three basic types of requirements. The first involves "interfaces" which, in turn, are divided into three sub-types: intra-organization interfaces, inter-organization interfaces, and external interfaces.

The second basic type of requirement for which work sheets are provided concerns data regarding regulations and policies. The job here is not to assemble a catalog of all such requirements, but rather to record only those sections or features that are pertinent to the system under study.

The final basic type work sheets which are provided relate to system-user requirements and preferences. Each of these basic requirement types is described more fully in the paragraphs that follow.

DETERMINING INTERFACE REQUIREMENTS

A client once asked, "What kind of face is an interface?" A useful definition, as it might apply here, would be that an interface is a confrontation between the source and user of information. This might be in the form of a meeting between two or more persons, or a written report, or it could involve the communication between two machines where, for example, a data processing tape generated by one system must be compatible with another.

Information relative to the sources and users of data is recorded on several interface requirements work sheets that are provided in the Requirements section of the Workbook. The job, here, is to determine and document all data required as input to the system and as outputs of the system, whether currently in written or oral form.

Intra-Organization Interfaces Work Sheet

Interface requirements within the organizational unit directly involved in the study are recorded on the Intra-Organizational Interfaces work sheet shown in Figure 31. This would include, of course, the types of information required both from an operational and management standpoint. In using the work sheet, the units or sub-units involved in each interface are noted, and the purpose is recorded. Interfaces can be in the form of information being submitted or reported via phone calls, teletype messages, meetings, written reports, data processing tapes, cards, or other media. The frequency of the interface is to be recorded, as well as the key information elements involved in the interface. Examples of key information elements might be "reporting daily attendance" or "report of weekly production." If the interface medium is a documented one, separate Document Identification work sheets should be filled out for each such document. This particular aspect of data collection is covered more thoroughly in a later chapter.

The purpose here, of course, is to isolate and identify all of the required information elements that must be exchanged within the organizational unit being studied, regardless of the medium of exchange.

Inter-Organization Interfaces Work Sheet

Interfaces required within the company but external to the department or unit under study, are to be recorded on the Inter-Organization Interfaces work sheet shown in Figure 32. This form, virtually identical to the Intra-Organization Interfaces work sheet just reviewed, concentrates on another important facet of interface requirements. These

INTRA-ORGANIZATION INTERFACES

Intra-Organizational interfaces are required within this organization between (sub-unit):

and

Purpose

Interface Medium (Phone, Teletype, Meeting, Written Report, Tapes, Cards, etc.)*

Frequency

Key Information Elements

and

Purpose

Interface Medium (Phone, Teletype, Meeting, Written Report, Tapes, Cards, etc.)*

Frequency

Key Information Elements

and

Purpose

Interface Medium (Phone, Teletype, Meeting, Written Report, Tapes, Cards, etc.)*

Frequency

Key Information Elements

*Prepare separate Document Identification work sheets where appropriate.

FIGURE 31: Intra-Organization Interfaces Work Sheet

Inter-Organizational interfaces are required between this unit:

____:

and ____

Purpose ____

____

Interface Medium (Phone, Teletype, Meeting, Written Report, Tapes, Cards, etc.)*

____

Frequency ____

Key Information Elements ____

____

____

and ____

Purpose ____

____

Interface Medium (Phone, Teletype, Meeting, Written Report, Tapes, Cards, etc.)*

____

Frequency ____

Key Information Elements ____

____

____

and ____

Purpose ____

____

Interface Medium (Phone, Teletype, Meeting, Written Report, Tapes, Cards, etc.)*

____

Frequency ____

Key Information Elements ____

____

____

*Prepare separate Document Identification work sheets where appropriate.

FIGURE 32: Inter-Organization Interfaces Work Sheet

include those types of information exchanges such as reports to higher levels of management, and administrative and financial type reports prepared by one organizational unit as a service for another. Again, as with all interfaces, the information that should be recorded on this work sheet involves the exchange of information regardless of medium. Also, as with Intra-Organization Interfaces, Document Identification work sheets should be prepared for all written reports and other such documents.

External Interfaces Work Sheet

Interface requirements relative to data sources or users external to the organization's company are to be recorded on the External Interfaces work sheet, Figure 33. As with the previous two interface work sheets, this form provides space for recording information relative to the organizations involved in each interface, the purpose of the interface, the medium employed, the frequency, and the key information elements involved. This important category involves such things as the preparation of reports or forms for municipal and government agencies (tax reports, for instance), and the exchange of necessary information between firms or agencies.

In collecting interface data, problems with the current method can often be discovered. A good example of this involved two large Federal agencies which, for the purpose of describing the problem, are designated here as agency A and agency B.

The systems analyst, in conducting interviews at agency A and recording information relative to external interface requirements, recorded the following requirement. A data processing tape, with information derived from an output of one of agency A's systems, was required for delivery to agency B each month. The tapes had to be compatible (machine readable) to an existing agency B data processing system. The two systems, themselves, were not compatible, so when the requirement had first been urgently imposed two years earlier, agency A management decided on an expedient, but temporary, solution. It was this. An employee would take the agency A output tape, have a printout produced, manually edit it to the agency B format, have it keypunched, and produce and deliver a tape with the same information but compatible to the agency B system. This process, at the time the systems analyst had uncovered it, had been going on several years occupying the full time of one employee. Agency A's management, occupied with new problems, had long ago forgotten their "temporary" solution. By discovering this problem it was easy enough to incorporate into the new-system design a feature that automatically produced a tape to the agency B format, this at a great savings in cost and time.

DETERMINING REGULATIONS AND POLICIES

The importance of policy and regulatory requirements depends largely on the organization under study. Heavily regulated enterprises and agencies are often severely restricted as to how they are permitted to perform certain functions. Furthermore, heavy reporting burdens are often imposed on them. Firms in the transportation industry would be included in this category, for example. Another field where systems of operation are acutely affected by regulations is the securities industry. In the stock clearing

External interfaces (outside the company) are required between this organization
____ :

and ____
Purpose ____
____

Interface Medium (Phone, Teletype, Meeting, Written Report, Tapes, Cards, etc.)*
____

Frequency ____
Key Information Elements ____
____
____

and ____
Purpose ____
____

Interface Medium (Phone, Teletype, Meeting, Written Report, Tapes, Cards, etc.)*
____

Frequency ____
Key Information Elements ____
____
____

and ____
Purpose ____
____

Interface Medium (Phone, Teletype, Meeting, Written Report, Tapes, Cards, etc.)*
____

Frequency ____
Key Information Elements ____
____
____

*Prepare separate Document Identification work sheets where appropriate.

FIGURE 33: External Interfaces Work Sheet

function, for instance, precise hours are specified when input must be entered into the system, and when certain types of reports must be generated.

Policies, especially those dictated by labor considerations, can also greatly affect a system's mode of operation. The information pertaining to these types of considerations should be recorded on the Regulations and Policies work sheets.

There's a fine line of distinction between what are considered regulations and policies, and what are considered operating procedures. For present purposes, operating procedures are considered to be formal documentations of the existing system and thus, subject to change to support cost- and benefit-proven features of a new or improved system. For that reason, procedures are discussed in the following chapter relative to the study of the existing system flow. Regulations and policies are considered here as inviolate and necessary, constituting requirements that the system *must* comply with.

The Regulations Work Sheet

Regulations affecting the system under study should be recorded on the Regulations work sheet, Figure 34. Space is provided for recording the name or title of the particular regulation, and identification number or code, and the source of the regulation in terms of who it was issued by and who authorized it.

The general purpose of the regulation should be briefly noted, and a listing of the organizational units that are affected by the regulation should be shown. If a copy of the complete regulation is not available to the systems analyst, then he should note where a complete set is located, and the status of the availability. The bottom of the work sheet provides space for recording excerpts or extracts of just those sections or features particularly pertinent to the system being studied.

Union agreements may also impose restrictions that affect the operation of the system. These, too, should also be recorded on the Regulations work sheet.

The Policies Work Sheet

Information as to policies that affect the system under study should be recorded on the Policies work sheet shown in Figure 35. Virtually identical in format to the Regulations work sheet, this form provides space for recording policy-type information generated by the organization's management. This might be information that reflects a management attitude resulting from either having or not having a union. It can impose restrictions on the way that personnel might be utilized in the operation of the system. As with regulations, the systems analyst should limit himself to recording only that data that has an impact on the system he is analyzing.

DETERMINING USER REQUIREMENTS AND PREFERENCES

Closely allied to interface requirements are user requirements and preferences. Since it is often difficult at first exposure to differentiate between system requirements and preferences, it is safer for the analyst to solicit all ideas as to system needs, even at the

REGULATIONS

Affecting System Under Study

(If more than one, use separate sheets to identify)

Regulation Name or Title ______________________________

____

Identification Number or Code ______________________________

Regulation Source:

Issued by ______________________________

Authorized by ______________________________

General Purpose ______________________________

____

____

____

____

Organization Unit(s) Affected:

____

____

____

____

____

____

Copy of Complete Regulation Located ______________________________

____

______________________________ Available?__________

Section(s) or Feature(s) Pertinent to System Under Study

____

____

____

____

____

____

____

____

____

____

____

FIGURE 34: Regulations Work Sheet

POLICIES

Affecting System Under Study

Policy Name or Title ______________________________

____

Identification Number or Code ______________________________

Policy Sources:

Issued by ______________________________

Authorized by ______________________________

General Purpose ______________________________

____

____

____

____

Organization Units(s) Affected

____

____

____

____

____

____

Copy of Complete Policy Located ______________________________

____

______________________________ Available? __________

Section(s) or Feature(s) Pertinent to System Under Study

____

____

____

____

____

____

____

____

____

____

____

____

____

FIGURE 35: Policies Work Sheet

risk of duplication. Requirements that are necessary can be segregated from desirable features later.

In soliciting this type of information through interviews, the analyst often finds interviewees extremely cooperative. It affords these system users the opportunity to offer ideas and suggestions that can greatly help them in their work, and improve the usefulness of the ultimate system.

The User Requirements and Preferences Work Sheet

Both necessary and desirable system features are to be recorded on the User Requirements and Preferences work sheet shown in Figure 36. The work sheet provides spaces for identifying the user's name, organizational unit, and title or position. Next, space is provided for noting the required or preferred system characteristic and the reason for the need. Whether or not the current system provides this particular characteristic should be noted and if not, the reason given why it is not provided.

The characteristics that a system user might want could involve faster information, more accuracy, or other such features. If the characteristic involves information, the source of such information, if known, should be specified.

It will be noted that the work sheet provides space for recording only one required or preferred characteristic. This is so that later the systems analyst can easily sort all such work sheets, merging them into various types of convenient groupings. He may find, for instance, ten separate suggestions regarding a single type of report. With separate work sheets these can all be merged together into one easy-to-evaluate packet.

The User Preference Checklist

Once the systems analyst has completed his interviews and has recorded required and preferred system characteristics on the appropriate work sheets, he must segregate them according to necessary requirements and desired features. The necessary requirements will be used in the development of the Statement of System Requirements, covered in the following section of this chapter. Desirable system characteristics are transcribed to the first column of the User Preferences Checklist shown in Figure 37. In this form the user preferences can go through several different stages of evaluation. One level of evaluation involves determining the cost of incorporating this characteristic into the new system and enumerating the benefits that can be derived. Another level of evaluation involves management's concurrence or rejection of the suggestions. If the results of either of these two evaluations for any given suggestion is negative, the reason for rejection is recorded in the right-hand column of the checklist. If the preferred characteristic is to be included in the new or improved system, a check mark indicates that fact. Later, when the new-system flowchart is developed, a cross-reference number is added to the checklist indicating where on the flowchart that particular characteristic is accommodated. In this fashion the client, later, in his final review of the proposed new or improved system, can see that system in relationship to what the users have asked of it.

An example of a portion of a filled out User Preferences Checklist is shown in Figure 38.

User (Individual's Name) ______________________

Organization Unit ______________________

Title or Position ______________________

Required or Preferred System Characteristic ______________________

Reason for Need ______________________

Does Current System Provide This? ______________________

If Not, Reason (if known) ______________________

If Characteristic Involves Information, Specify source (if known) ______________________

FIGURE 36: User Requirements and Preferences Work Sheet

USER PREFERENCES
Checklist

Sheet__of ____

Preferred Characteristic	Check (✓) If To Be Included In System	Flowchart Ref.No.	Comment If Not Included In System

FIGURE 37: User Preferences Checklist

USER PREFERENCES
Checklist

5.7
Sheet 4 of 12

Preferred Characteristic	Check (✓) If To Be Included In System	Flowchart Ref. No.	Comment If Not Included In System
4.2 PROJECT OFFICER SHOULD	✓	P-176	
RECEIVE PRINTOUTS OF			
DISAPPROVALS IN ORDER TO			
COMPLETE PROJECT FILE			
4.3 A MONTHLY REPORT SHOULD			SPECIAL REPORT MAY
BE GENERATED ON THE			BE GENERATED BUT
FREQUENCY OF USAGE OF			THIS WILL NOT BE
TERMS USED TO INDEX			STANDARD REPORT
PROJECTS			
4.4 EACH GROUP SHOULD	✓	P-238	
RECEIVE COMPLETE			
FISCAL LIST			

FIGURE 38: Portion of a Completed User Preferences Checklist

DRAFTING THE STATEMENT OF SYSTEM REQUIREMENTS

Each systems analysis project starts with a Statement of System Objectives. At a later point in the analysis, once all necessary system requirements have been determined, and the existing-system flow has been charted, those original objectives are translated into a statement of System Requirements. Input for this includes the original objectives, and the work sheets specifying interface requirements, regulations and policies, and user requirements. An example of a portion of a completed requirements statement is shown in Figure 39.

The first entry ("14.0 SUBMIT VOUCHERS") is a general statement of a function performed during operation of the system. This statement would normally be derived from one or several of the various requirements work sheets and if it was a requirement being complied with by the existing system, it would also be reflected in the description of the existing system. Since all information handled by this system, to be useful, supports an operation or a decision, requirements will necessarily be related directly to such a function. Then this approach provides a convenient and systematic way of enumerating and sorting requirements, and also provides a test of their validity.

The ultimate format of such a statement of system requirements may differ from that which is shown in the illustration. This depends, of course, upon the uniqueness of the particular system which is being analyzed. Essentially though, regardless of format, the statement should contain every single requirement imposed on the system.

STATEMENT OF SYSTEM REQUIREMENTS

Sheet___of___

14.0 SUBMIT VOUCHERS

At designated times during and at conclusion of contract, grantee submits vouchers for contractually covered expenses. These are the basis of expenditure of obligated funds.

Information Required

Allowable expenses (contained in contract); Previous approved expenditures and funds remaining.

SYSTEM FUNCTIONAL REQUIREMENT

14.1 Capability of producing statement of funds contracted, expended, and balance.

Information Generated

Funds Expended by grantee by cost categcry.

SYSTEM FUNCTIONAL REQUIREMENT

14.2 System must be able to accept data direct from voucher by keypunch or other means.

FIGURE 39: Example of Portion of a Statement of System Requirements

Once this final Statement of System Requirements has been drafted, as with the User Preferences Checklist, it should be reviewed by the client-management. Once approved it becomes one of the main specifications for the design of the new or improved system. As with the User Preferences Checklist, the requirements will be ultimately cross-referenced with the new-system flowchart so that the client can easily check and see that requirements have, indeed, been met. Examples of how this cross-referencing can be accomplished are shown in a later chapter.

6
Efficiently Documenting Existing-System Operations

The previous chapter was concerned with defining requirements to which the system must conform. This chapter deals with a parallel activity, that of documenting how the existing system works. One important part of this documentation, that of constructing an existing-system flowchart, is significant enough in importance to warrant a chapter of its own. A chapter on documenting existing-system data completes the system definition. There is also a separate chapter concerned with discovering why the system operates the way it does and what its current costs and benefits are. Together these chapters describe the key tasks of defining and analyzing a system preparatory to planning improvements. They provide the base line from which these improvements can be designed and implemented, the ultimate goal of the project.

In establishing how the existing system operates the analyst is concerned with four categories of data:

- Information Flow
- Operations Performed
- Data Content and Format
- Equipment and Organizations Utilized

The flow of information, or system flow, as the term is used here, means the transmission of data from one individual, organization, document, or file to another. If we say that a document is sent from the Receiving Department to the Accounting Department with a copy to Production Planning, we are describing a portion of the flow, specifically two separate transmittals. System flow is best "described" and easiest understood when presented graphically on a flowchart.

Operations performed include anything that is done to, or with, the data. They may include the preparation of a report, the recording of information on a form (or keypunching onto cards), or the storage of a document in a file. Operations might also include calculations or other manipulations that transform the data, or that combine or merge two sets of data into a single file, or conversely, sort the data of one file into several others. The latter type of operations need not be automated sorts and merges; these operations might also be done manually. Decisions, analyses, and evaluations are other categories of operations that are of extreme importance.

SOURCES OF INFORMATION

The information that describes an existing system is derived primarily from interviews, examination of procedures and system documents, and from direct observation of the system. Much of the information will be developed during the interviews by direct questioning, by asking respondents what they do with information, where they get it, and how they distribute the information they generate. A good deal of every interview will thus be devoted to discussions of system flow and operations. Very often this will center around the handling of a particular form. Detailed questioning about the way a form is handled is a very convenient way of finding out about both system flow and operations. The processing of a form is also something the respondent can relate to much more easily than specific questions about "flow," a term that could be meaningless to him. For example, the analyst might be told, "We receive a copy of the packing slip from Receiving as soon as the shipment comes in (a transmittal, an element of flow). The clerk matches it to the invoice (an operation) which came in by mail (flow), records this in her journal (a second operation), and posts the amount to Accounts Payable (a third operation). Under some circumstances she prepares a voucher (a decision and an operation—at this point the analyst should determine what information is used as a criterion for that decision) and then she files the copies." The latter step probably involves a "sort" of some type, perhaps alphabetically by vendor name, and a transmittal to the file.

In any of these interview situations the respondent is asked how he does his job, on what basis must decisions be made, what specific information he requires, and where he gets that information. The systems analyst should also find out how the respondent indicates his approval or disapproval when called for, whether reports are prepared as a result of his actions, where they are transmitted, and other similar data. The analyst should, of course, find out exactly what forms and documents are involved, and secure copies. In the case of forms, especially, it is useful in their later analysis if they are completed (filled out). If any portions of these are ambiguous or confusing, clarification should be obtained at the time of the interview.

Additional data will be obtained by an analysis of the formal procedures used by the respondent or his department. Procedures, as has been pointed out, are the "programs" governing manual systems. They specify operations to be performed and direct how they are done. They also indicate what transmittals are to be made and how.

If any parts of the process are automated, data on the programs used must be gathered. These will not generally be obtained from the same people that provide data on procedures and manual operations. Information on programs and automated systems may be obtained from the data processing department, or in larger organizations, the staff analysts. They may even be documented in a form that can be useful to the system analyst.

Although the flow of information and the operations performed must be separated intellectually when analyzing the system, they obviously cannot be separated during the data collection process. Both operations and flow are interspersed throughout the formal procedures; both will also be commingled during the interview. The work sheets, which are described in this chapter, therefore, are designed for recording both types of information on the same form. They will support the preparation of the graphic flowchart and also serve as primary documentation of the details of existing operations.

Information on the equipment and hardware used will come from a variety of sources. Some will be obtained in the initial phases of the project, some during the interviews, and a good deal from the equipment literature and system documentation itself. Work sheets for recording this data were described in a previous chapter. In documenting the existing system, it is important to identify specific machines with specific processes. It is also important to keep checking the equipment work sheets when a piece of hardware is referred to in order to ensure that these work sheets are complete.

The same is true of the organization and its relationship to the system description. It is essential in analyzing the system to know which organizational units do what. The organizations responsible for an operation or for the transmittal or receipt of data must be identified during the interview and keyed to these elements when they are recorded on the work sheets.

DOCUMENTING SYSTEM FLOW AND OPERATIONS

There are two major sections of the Workbook that are designed to define and document how the existing system operates. These are the Existing-System Flow and Existing-System Documents. The work sheets in the first deal primarily with system flow and operations. As has been pointed out, they support the preparation of the system flowchart. In addition, they provide for making notes as to system features susceptible to improvement as these ideas evolve. The work sheets in the second section deal with the forms, reports, and files used in the system. Although they are of some use in charting the flow, they are basically used to support the analysis of the data itself.

The work sheets covered in this chapter include the following:

- System Flow
- System Features Susceptible to Improvement
- Procedure Summary
- Automated Systems Identification
- Automated Systems Breakdown
- Automated Systems Summary
- Developmental Program Status

System Flow Work Sheet

The job is to secure a narrative description of the existing-system flow. In documenting this description, it should be in a form from which the flowchart can easily be constructed. Later, when the flowchart has been completed, it should be appended to these work sheets and inserted into the Workbook.

The System Flow work sheet, shown in Figure 40, has been designed for just this purpose. It is provided in the Existing-System Flow section of the Workbook. The work

SYSTEM FLOW **3.1**

System Element________ Sheet___ of ___

Prime Org. Unit________ Governing Procedure________

Seq.	Process	Con	Input/Output	Con	File/Store	Con	Decision	Con

FIGURE 40: System Flow Work Sheet

sheet should be used to document both formal and informal procedures. It can be used to document either automated or manual processes or combinations of the two. It can also be used to trace the processing of a particular form or document through an organization as was discussed earlier in the chapter.

At the top of the work sheet are spaces for identification data which can be used to key this to other processes or related documents. The system element described is to be identified. It should be captioned in such a manner so as to easily relate it to other system elements.

If there is a formal procedure document that directs how this process is to be done, the name and number should be entered after "Governing Procedure." If this is an informal procedure, a notation should be made to that effect, and as to the status of the informal procedure. It may be a very stable process that has been done the same way for many years but has never been formalized. On the other hand it may be something that the manager or supervisor of this group is experimenting with. If he is meeting resistance this is important to the analyst and must be noted and described (on a supplemental sheet if necessary).

Another category of undocumented process that is important to know about comprises the unofficial modifications or deviations from an established and formal procedure. These may include short cuts which may be peculiar to one group or may be commonly practiced. Conversely, they may be expansions of an inadequate procedure. In this case the formal procedure should be identified with a notation that this work sheet reflects a modification. The steps should also be identified as to steps that are in the procedure and steps that take the place of, or are added to, the documented steps. A reason for recording these deviations is that they are very often symptoms of inadequacies in the existing system and will provide clues for planning system improvements. The essential thing for the analyst to remember, though, is that the system should be described in terms of the way it actually operates, formally or not, with or without sanction.

The system element being described on the work sheet should be identified to some prime organization. This might be the organizational unit that is involved in most of the steps of the process, or is basically responsible for seeing that this portion of work is accomplished.

As to the main body of the work sheet, it is designed for recording all of the basic elements of a system flow. The left-hand column provides space for the analyst to assign a numbered sequence to each step in the flow. These numbers enable him to easily cross-reference connections between process steps, forms, files, and decisions. The "Process" column itself is where the analyst describes each separate processing step. If input or output material, or files or storage, or a decision is associated with that process, it is so noted in the appropriate column. If any of these notations connect to any other process other than the one following next in sequence, then a cross-reference sequence number is noted in the "Con." (Connector) column. This can be used to connect to a future process, or to connect to a past process for steps that must be re-cycled.

It is easiest to understand the use of the System Flow work sheet by showing how it may be used in actual practice. In this example, the system flow is being described in an interview, the analyst recording the information on a work sheet. (This work sheet is shown later in this chapter as Figure 41.) It should be remembered that the analyst's job is to get a description of the system as it is, not as it should be. The interview dialogue might go something like this:

Analyst:	Now, I'd like for you to describe to me the way you receive and store goods.
Interviewee:	Well, as soon as the goods get here we check, first, to see if they were ordered.
Analyst:	Who checks?
Interviewee:	The Receiving Clerk.
Analyst:	Does he have copies of the original orders?
Interviewee:	Yes, he has a file cabinet with all the current open orders in it.
Analyst:	So he checks the shipment against the open orders. Then what?
Interviewee:	Well, if we didn't order it he sends the whole thing back.
Analyst:	Does he record this in any way?
Interviewee:	You mean, make a report or something?
Analyst:	Yes. Or maybe check to see if his open-order file is complete.
Interviewee:	No. All he does is mark on the packing slip that we didn't order these goods, then ships the whole thing back.
Analyst:	Let's say the goods were ordered. What next?
Interviewee:	Well, he checks the goods against the packing slip to see if everything's there. Any "shorts" or back-orders, or things like that.
Analyst:	And if there are shortages?
Interviewee:	He fills out a "Receiving Discrepancy" notice and sends it to the Accounts Payable department.
Analyst:	Does he keep a copy?
Interviewee:	No. But he makes a note about *all* goods received on the original orders.
Analyst:	What does he do with the packing slips?
Interviewee:	He sends them up to Accounts Payable, too.
Analyst:	What do they do with this material?
Interviewee:	I don't know. You'll have to ask them about that. (Analyst puts a check mark on work sheet after the description of this last process, indicating to himself a branch of the flow he'll have to pick up from another source)
Analyst:	What does the Receiving Clerk do with the goods, now?
Interviewee:	Well, we have a file of storage instructions for each kind of goods we receive. He gets one of those. Then he sends it over to the stockroom. The clerk there checks to see if the instructions are with the goods.
Analyst:	Why doesn't he have the file of instructions there in the stockroom?
Interviewee:	I don't know.
Analyst:	What happens if the instructions are not with the goods?
Interviewee:	He sends the whole thing back to Receiving.
Analyst:	The goods?
Interviewee:	Yes.
Analyst:	And if everything's in order?
Interviewee:	He gets out his Stock Code Catalog, locates the right number, and identifies the goods with it. Then he stores them.
Analyst:	Does he have an inventory system?
Interviewee:	Yes. That's all procedurized. It's procedure number S-21. I can get you a copy of that.
Analyst:	Yes, I'll need that. Now, what happens to the storage instructions when he's through with them?
Interviewee:	He sends them back to Receiving. They use them over again.

As the foregoing interview takes place, the analyst records the information on the System Flow work sheet. Possibly he records very cryptic notes during the course of the interview, refining them as soon as possible after the interview has terminated. The way the final work sheet might look is shown in the example shown in Figure 41. How different analysts will record data depends on their personal preferences. Whether "goods" should be shown as input or not, for example, is up to the analyst. It depends on whether or not he feels it ultimately lends clarity or not to the system description.

The technique for translating the information recorded on the system flow work sheets into flowchart form is described in Chapter 8. A flowchart version of the just-

SYSTEM FLOW **3.1**

System Element RECEIVING & STORING GOODS Sheet 1 of 1

Prime Org. Unit ______ Governing Procedure ______

Seq.	Process	Con	Input/Output	Con	File/Store	Con	Decision	Con
1	RECEIVING CLERK RECEIVES GDS		GOODS					
			PKG SLIP					
2	COMPARES AGAINST ORIG. ORDER		ORDER		OPEN ORDER FILE CAB			
3	ON ORDER ?						NO	4
							YES	5
4	NO. MARKS PKG SLIP "NOT		GOODS					
	ORDERED" & RETURNS TO SHIPR		PKG SLIP					
5	YES. CHECKS GOODS AGAINST		GOODS					
	PACKING SLIP		PKG SLIP					
6	MATCH ?						NO	7
							YES	8
7	NO. PREPARES RECEIVING		REC. DISCR.	8				
	DISCREPENCY NOTICE		NOTICE					
8	YES. NOTES REC'D GOODS &							
	DESCREPENCIES ON ORIG ORDER							
9	TRANSMITS DATA TO ACCT'G	✓	PKG SLIP					
	DEPARTMENT		DISC. NOTICE					
10	SECURES STORAGE INSTRUCT'NS		STOR. INST.		STOR. INSTR. FILE			
11	TRANSMITS ALL TO STOCKROOM		GOODS					
			STOR. INST.					
12	STOCK CLERK RECEIVES		GOODS					
			STOR. INST.					
13	STORAGE INSTRUCTIONS		STOR. INST				NO	14
	WITH GOODS ?						YES	15
14	NO. RETURNS GOODS TO REC'G		GOODS	10				
15	YES. DETERMINES STOCK CODE NO.		STK CODE CAT					
16	ID'S GOODS WITH CODE NO.							
17	UPDATES STOCK INVENTORY	S-21						
	SYSTEM (SUB-ROUTINE)							
18	STORE GOODS - CYCLE COMPLETE		GOODS		GOODS STORE			
19	RETURN STOR. INST. TO REC'G		STOR. INST.					
20	REC. CLK RE-FILES STOR. INST.		STOR. INST.		STOR. INST. FILE			

FIGURE 41: Example of a Completed System Flow Work Sheet

described example work sheet is shown in that chapter in Figure 63. Once such flowcharts are completed they should be appended to the system flow work sheets and inserted into the Workbook.

Checklist for System Features Susceptible to Improvement

Throughout the entire analysis portion of a project, the systems analyst is continually confronted with ideas for improving the system. Some of these are in the form of suggestions from system users. Others emerge as a result of documenting the system as it is, or from studying and analyzing the various system elements.

In the example just given regarding the use of the System Flow work sheet, several possibilities for system improvements may have occurred to the analyst as he recorded the data. One idea might be to have the Receiving Clerk, when unable to locate an open-order in his file, make sure that the order is not somewhere else in the system before going to the trouble of returning the goods to the shipper. Another idea might be to locate the Storage Instruction file in the stockroom rather than the Receiving area, eliminating several possible back-and-forth transmittals of goods and documents.

Even though suggestions for possible improvements like these may be obvious in a study of the work sheets themselves, it is useful to extract these ideas, as they occur, from the System Flow and other work sheets, and summarize them in one place. The form provided for this purpose is titled System Features Susceptible to Improvement and is located in the Existing-System Flow section of the Workbook. Improvement suggestions are noted while they are fresh in the analyst's mind. Later he can use this as a checklist he, himself, has created. An example of how it can be used is shown in Figure 42. As ideas are recorded, the analyst can indicate with a check mark whether each improvement possibility might be included in the new system design or taken care of more directly as a "quick fix." These are items he can discuss with the department head at his next review.

SYSTEM FEATURES SUSCEPTIBLE TO IMPROVEMENT **3.2**

Checklist

System Feature	Check Implementation Possibility (✓)	
	Quick Fix	New System
1. IN RECEIVING & STORING GOODS, RECEIVING	✓	
CLERK SHOULD MAKE SURE AN ORDER IS NOT		
SOMEWHERE ELSE IN SYSTEM BEFORE RETURNING		
GOODS TO SHIPPER WHEN HE CAN'T FIND AN		
OPEN ORDER IN HIS FILE.		
2. IN RECEIVING & STORING GOODS, THE STORAGE	✓	
INSTRUCTION FILE SHOULD BE LOCATED IN THE		
STOCKROOM RATHER THAN IN THE REC-		
EIVING AREA.		

FIGURE 42: Example of Use of the System Features Susceptible to Improvement Checklist

Procedure Summary Work Sheet

Some organizations operate without formal procedures; others are highly procedurized. In those situations where procedures do exist the systems analyst should make an inventory of procedures, listing those that affect the system being studied on the Procedure Summary work sheet shown in Figure 43.

As previously described, preparatory to developing the flowchart of the existing system, it is sometimes useful to translate procedures to flow terms by using the System Flow work sheet. This practice, though, is not necessary in all cases. The procedure, as

PROCEDURE SUMMARY 3.3

Procedure Name & General Subject	I.D. Code or Number	Copy Located	Translate To System Flow Work Sheet?

FIGURE 43: Procedure Summary Work Sheet

written, may already be in a form useful for flowcharting, for instance. Or the process itself may not be of sufficient importance to warrant this type of effort.

The Procedure Summary work sheet, then, is for listing all related procedures and indicating (in the right-hand column) which ones should be translated to a System Flow work sheet. The work sheet provides space for noting the procedure name and general subject area, its identification number, and the location where a complete procedure may be obtained. This work sheet provides the analyst with a complete inventory of information he needs to know about procedures at the time he's ready to draft the existing-system flowchart.

Automated System Work Sheets

For those systems being studied that are already automated to some degree, special work sheets are provided in the Existing-System Flow section of the Workbook. Although the previously described System Flow work sheet can be used to describe operations performed by the automated portions of a system, additional data would be needed about the existing programs. Sometimes these programs are already documented completely with full descriptions of the systems, detailed sets of program flowcharts, source decks, and program listings. In cases like these, and if new changes are in process, the analyst merely needs to obtain a set of the documentation, and work sheets may not be needed. At the most only a few of the items called for need be recorded.

If the existing system is highly automated, an Automated System Summary work sheet will be desirable even if the programs are well documented. This work sheet will be discussed later in the chapter and parallels the Procedure Summary work sheet.

There are three other basic work sheets included for recording automated system data. The first two are used to identify and describe operational programs. The Automated System Identification work sheet is used to provide technical data on broad applications of programs to particular system operations. The Automated System Breakdown provides for data on specific programs or program modules. The third work sheet is used to record data on any programs under development or program modifications in process that will have a significant effect on the total system under study.

Automated System Identification Work Sheet. This work sheet is illustrated in Figure 44. Much of the information recorded on this work sheet will not be directly related to system flow or data analysis but will be required later in analyzing the costs and benefits and in establishing criteria for evaluating proposed changes. It is recorded on this separate work sheet for convenience in collection since all this type of information will generally come from the data processing personnel rather than from interviews with the operating or user groups.

The Automated System Identification work sheet is intended for the broad category of software categorized as applications programs, or programs used to perform particular system operations, rather than utility or executive software. These "applications packages" may be developed especially for the existing system, or they may be purchased, or "general purpose" packages.

It may be difficult to interpret detailed programming data in these terms, particularly if the system used a highly modular approach to programming or utilizes time-

AUTOMATED SYSTEM IDENTIFICATION 3.4

Name ______

Other Identifiers (Codes, etc.) ______

Purpose ______

______

______

______

Batch? ______ On-Line? ______ Time Sharing? ______ Other? ______

Core Storage Required ______

Central Processing Unit (CPU) Used ______

CPU Time Required ______

System Environment ______

______

Auxiliary/Peripheral Equipment Required (Include teleprocessing devices and terminals, if applicable):

______ ______

______ ______

______ ______

Program Language ______

Program Operational? ______

Purchased? ______ Leased? ______

Who Developed? ______

Who Maintains? ______

What Provisions for Modifications? ______

______

______

Documentation Available ______

______

From Whom? ______

Description (Inputs/Outputs, Data Bases, Modules, Functions, etc. Use supplemental sheets, as required)

______

______

______

______

______

FIGURE 44: Automated System Identification Work Sheet

sharing services. However the system analyst at this stage should orient himself to the total system and the set of operations that must be performed. He should step back from the details of the specific programming techniques and look at a set of programs in terms of the system operations. That is, he should analyze the programs from a user's point of view rather than from that of the computer operator or programmer. When it is necessary to also record the programming details, the Automated System Breakdown work sheet, which will be discussed later in the chapter, can be used.

The first lines of the work sheet are used for identifying the system and programs.

This data should be as precise as possible to facilitate later reference and for identifying the application itself to the program documentation.

A brief summary statement of the overall function and usage of the programs should be entered in the space designated for "Purpose." This should only be a summary, to be elaborated further in the space designated for description. Sometimes the title of the program or system will be sufficient to describe the purpose of the software.

Next on the work sheet are spaces for indicating the operational mode of this system, that is, whether it is used in the batch or on-line mode, or whether or not it operates on a time-sharing basis. Other modes can also be entered, either in place of these three or to supplement them. For example, it may be a remote batch-entry system or, rarely, a real-time system. It also may be a batch system but run at an external data processing service bureau which could be a significant fact later in the evaluation of costs.

Memory required for the programs should be recorded to indicate the size of the programs and for later use in evaluating costs. If storage requirements are not available, some indication of program size should be noted.

The type of central processing unit used and the time used during a typical processing cycle should be entered in the spaces provided. The time may be rather difficult to establish or to define properly, but is vital to evaluating costs of system operation. It should be recorded in whatever units are available and these units well defined. Usually it can be described in terms of minutes for so many transactions or for the processing of a certain number of records. The reason for precisely defining the measurement units is that "transactions," for instance, may mean one thing to the user and another to the computer operator. For example, to the Payroll Department, a single transaction might be the computation of accrued wages for a single employee. To the computer operator, on the other hand, computing one person's pay might mean three or four transactions if several modules are used. Records are also variable, as for example "physical" versus "logical" records, and must be defined.

The way time is measured in the computer room should also be indicated since the difference between "wall clock" time and metered time can be very significant. For metered time, the formula used is also of importance and should be indicated on the work sheet.

Very often machine usage reports and other similar statistical reports are provided to the computer center manager or to the user. If access is available to these, and they are sufficient for the purposes of the analysis (with particular concern for estimating and comparing costs), copies can be attached to the work sheet and a reference to them can be noted, thus avoiding the complexities of defining the run time required.

CPU time required and other variables are also highly dependent on the environment under which the programs are run. In the space designated "System Environment" should be indicated, for example, whether the system operates in a multi-processing, multi-programming, or other mode data on partitioning, further details on time-sharing parameters, or other applicable facts about the automated system environment.

The remaining spaces in this portion of the work sheet are for recording what auxiliary or peripheral devices are required for the system and what program language is used.

Following this section on the technical details of the program package are a group of spaces for recording information as to the software's origin and current maintenance responsibility. It will be important later in the new-system design phase to know whether the programs were developed internally, or purchased as an off-the-shelf package, or if they are being leased. In any case the possibility of modifying these programs may be of interest. For internally developed programs the identification of the group responsible for maintenance will then be of importance, and possibly it will be desirable to trace the programs back to the person or group which developed them in the first place. In most lease or purchase agreements there is provision for modification and maintenance either by the recipient or the vendor. These provisions vary widely and therefore should be noted in as much detail as possible, using supplemental sheets as needed.

The level and quality of program or system documentation also varies widely. Whatever exists, its availability will be of vital importance to the analyst. This data should be recorded in the space provided. As mentioned earlier, if good documentation is available, many of the spaces provided on this work sheet need not be completed. A reference to the appropriate documentation will be sufficient.

The lower part of the form provides space for a description of the system and for its current applications. The content of the description section will depend on the type of system being described. For a batch mode "application package," the standard description of inputs, outputs, data bases used, and functions performed will be required. In addition, data on the configuration of records will be useful. Both the content and format of inputs and outputs should be specified, as well as the media. In addition to specifying the content, inputs should be related to physical documents and keyed to Document Identification work sheets whenever possible. If the input is direct there may already be an Interface work sheet completed. If so, or if it would be appropriate to complete one, the information on that work sheet need not be duplicated. If the input is received from some other system in machine readable form, a precise identification of the tape and originating system should be provided to facilitate indicating this transmittal on the flow chart. Again, this could well be the subject of an existing Interface work sheet.

By the same token any printed outputs should be defined by report number or other formal identification. Machine readable outputs such as tapes, card decks, or disc packs should also be precisely identified and a notation made as to their location and usage. The same admonition about duplicating Interface work sheets or Document Identification work sheets applies. A reference to the appropriate documentation will be sufficient.

For a modular system, the names and functions of the key modules would be recorded and any significant information on how the system operates not included under "System Environment" noted. The Automated System Breakdown work sheet can be used for more detailed functional information, if this is required.

The specific data entered in this "Description" section will depend on the judgement of the analyst or the dictates of the project leader. The essential thing to remember is that the purpose of this work sheet is to support the description of the present system, to document data needed later to determine present costs, and to provide for benefit comparisons with proposed system changes. Programming or operating details which are not essential to these three functions, or which are contained in easily accessible program documentation, should not be recorded.

Automated System Breakdown Work Sheet. This work sheet, illustrated in Figure

AUTOMATED SYSTEM BREAKDOWN **3.5**

Name __________ I. D. No. /Code __________
System __________
System Flow Work Sheet Reference __________
Purpose/Usage/Description __________

Documentation __________

Name __________ I. D. No. /Code __________
System __________
System Flow Work Sheet Reference __________
Purpose/Usage/Description __________

Documentation __________

Name __________ I. D. No. /Code __________
System __________
System Flow Work Sheet Reference __________
Purpose/Usage/Description __________

Documentation __________

FIGURE 45: Automated System Breakdown Work Sheet

45, can be used primarily when the existing system uses modularized programs, such as commonly in use with time-sharing and multi-processing systems. It can also be used for recording information on significant utility programs, executive or supervisory compilers and assemblers, or specialized applications programs (as for example, photocomposition or Computer-Output-Microfilm packages), if this data is significant in accomplishing project objectives. Finally it might be used for engineering, scientific, statistical analysis, or other computational programs which do not need the detailed documentation in the Workbook required of programs related more directly to system operations.

For modularized programs, a separate block on the work sheet may be completed for

3.6

AUTOMATED SYSTEM SUMMARY

Program/Module Name & General Subject Area	I.D. Code or Number	Documentation Located	Other Work Sheets

FIGURE 46: Automated System Summary Work Sheet

each module, or only key modules might be described and a list attached of other modules (or reference made to the existing documentation). In the description block, data on parameter lists, function codes, and the like may be noted if they are significant to the total system flow. All such information which will be useful later in evaluating and comparing programs should be recorded.

Automated System Summary Work Sheet. The use of this work sheet, illustrated in Figure 46, is similar to the use of the Procedure Summary work sheet. It provides for listing programs, program packages, or program modules, with summary information about them whether or not they are the subject of a separate work sheet. Using the

Automated System Summary will be beneficial when the existing system is highly automated and utilizes a number of program packages, or when the programming system used is highly modularized but a separate work sheet is not justified for each package or module.

The work sheet provides for listing the programs or modules by name, indicating their identifying number or acronym, and briefly defining their purpose or function. Another column allows for entering the location where documentation can be obtained. There is also a column for indicating whether or not there are other work sheets perpared (such as System Flow work sheets, for example) that describe this particular unit.

DEVELOPMENTAL PROGRAM STATUS **3.7**

Name ______________________ I.D. No/Code ____________

System ______________________

Scope of Development (New programs, major modification, etc.) ______

____

____

____

Planned or In Process? ______________________

Purpose/Function of Programs ______________________

____

____

____

____

Reason for Development ______________________

____

____

Who is Developing? ______________________

____

Schedule for Becoming Operational ______________________

____

Anticipated Impact on Existing System ______________________

____

____

____

____

____

____

____

____

____

____

____

FIGURE 47: Developmental Program Status Work Sheet

As with the Procedure Summary, this provides the analyst with an inventory of information available both when he is preparing the existing-system flowchart and when he is analyzing or designing system changes.

Developmental Program Status Work Sheet. It is not an unheard-of phenomenon for a systems analyst to find he has been assigned to improve a system that is already in some stage of being improved. For this reason it is important for him to know the status of programs currently being modified, and of developmental work under way or planned. This data is recorded on the Developmental Program Status work sheet, shown in Figure 47. This work sheet is used whether the programming or other system change is simply planned or is already in work.

Following the identifying data is a space designated "Scope." Information as to whether completely new programs are being developed, or simply modifications being made to existing programs, should be recorded here. The extent of the modifications should also be noted and any other details which will help to measure the significance and magnitude of the developmental efforts.

Under "Purpose/Functions" a description of the programs themselves should be recorded. This information should define just what the developed programs are to do when completed, and the intended applications.

The problem which it is hoped will be solved or the objective of the development work should be recorded in the space designated "Reason for Developing." This information should include an indication of whether it is a data processing problem that is being solved or a system problem.

Space is also provided for identifying the group or individuals actually doing the development work, or performing the programming, and for the data or schedule established for making the programs operational.

In the final section, "Anticipated Effect on Existing System," the analyst should record his own analysis of the way this developmental effort will affect the total system. Anything that will be significant to the project or to the design of a new system should be recorded.

7
Describing and Analyzing Existing-System Documents and Files

The Existing-System Documents section of the Workbook provides work sheets for describing and analyzing the information that is processed by the system. Earlier chapters, when dealing with defining requirements and system flow, specified the use of document and file identification work sheets for describing these items as they became known to the systems analyst. This chapter describes these particular work sheets in detail, and outlines a method for analyzing the data elements.

The basic sources of information on data format are the physical documents themselves. As has been repeated several times before, it is important to obtain copies of all the documents produced or processed by the system. Although the work sheets, as shall be seen, provide spaces for precisely recording the specifics of the data, this is no substitute for a copy of the document itself.

The content of documents in which the data is recorded and the format of this data change as the information flows through the system. Each operation changes one or both of these attributes of data. It is important to record these changes in order to track information through the system, thus confirming the validity of a requirement or form. Recording the changes also further defines the operations performed. It is also essential in an automated process to precisely define the format of the data because of the dependence of programs, allocation of storage, and types of equipment on this factor.

There are several other reasons for documenting system information in detail. When systems changes are being evaluated, it will be important to know that a particular element, say an account or ledger number or a part number, is limited to five or eight characters because of the nature of the programs used to process it. It may also be important to know, as another example, whether a manual file contains certain historical information needed to institute a statistical report on trends.

The type of programming and equipment that will be required in automating a part of the system, or in modifying an already automated operation, can be drastically affected by the type of data to be processed. It will make a big difference, for example, whether the data elements are of fixed length and configuration, or can be forced into this format, or are of variable length. Conversely it may be found that a variable length data element could be reduced to a set of fixed-length codes. This could mean considerable savings in processing time.

Sometimes analysts will find circumstances where data is input to the system but never used. It will be processed through several steps, transformed in various ways, only to end up in a file and never referred to. This type of situation can be very difficult to determine without the careful tracking of data through the system.

Discretion must be exercised by the analyst, though, in terms of the extent of his analysis of data. The danger in attempting to track data and analyze their transformations lies in becoming bogged down in complexities and details. Some government agencies have literally spent millions in unsuccessful attempts to analyze and control data flow at the data element level.

The analyst is cautioned, therefore, to use the work sheets only to that level of detail that is useful to the requirements of his particular project. In this chapter five work sheets are presented. They are:

- Existing-System Document Identification
- Existing-System Manual File Description
- Existing-System Data Base Description
- Data Element Inventory Work Sheet
- Data Element Matrix

The first four of these work sheets are for recording basic identification and descriptive data. The matrix is useful in an analysis of the elements of data used in the system with an eye toward combining and simplifying system documents.

EXISTING-SYSTEM DOCUMENT IDENTIFICATION WORK SHEET

A "document" is normally thought of as printed or written on paper, a "hard copy." In automated systems, however, information may be transmitted from one set of processes to another without ever being printed out in hard-copy form. It is convenient to consider these transmittals also as documents, whether on tape reels, disc packs, card decks, or other modes, to insure complete coverage of the data throughout the system.

Another form of transmittal that can, for convenience's sake, be considered a document under certain conditions is the oral transmission of data from person to person. This type of transmittal has already been mentioned in connection with the Interface work sheets in the chapter where system requirements were defined. Generally, documentation on the Interface work sheets will be sufficient. However, if the transmittal is highly formalized and repeated in the same sequence in each transmittal, it may be possible in the design of the new system to substitute an automated transmittal. Because of this possibility, it is useful to also record this type of information on an identifying work sheet. An example of this type of condition might be the periodic reading of a set of gauges and meters by an oil-field worker which he "talks" back to a central station for recording.

When the document itself is available and readable, the Existing-System Document

EXISTING-SYSTEM DOCUMENT IDENTIFICATION 4.1

(Attach Filled-Out Example)

Flow Sequence or Ref. No. ________

Document Title ________

Identification Number/Code ________

Purpose or Function ________

Authority for (Procedure, Regulation, etc.) ________

System or Organization Preparing ________

Review Activity ________

Distribution and Number of Copies ________

Frequency of Issue ________

Annual Quantity ________

Media ________

Method of Reproduction (Blank Form) ________ Quantity ________

Method of Preparation (Completed Document) ________

Size and Number of Sheets ________

Document Contents (Attach completed sample and instructions for preparing. If sample not obtainable, record data elements)

Data Element Title	Size	Source

FIGURE 48: Existing-System Document Identification Work Sheet

Identification work sheet should be used as a "cover sheet" and filled out only to the extent that required data is not apparent from the document itself.

The Existing-System Document Identification work sheet is shown in Figure 48. Space is provided at the top of the form for the usual title and control or identifying number. The upper right-hand corner provides space for cross-referencing the document to the system flow. Under "Purpose or Function," a brief statement of the reason for having such a document, and its use, should be entered.

The next two spaces are important in establishing the formal requirement that resulted in creation of the document. The "Authority For" space should indicate the procedure or regulation that established this requirement. Normally, procedures will also describe the use of the document and, perhaps, give step by step directions for preparing it. These procedures may also be a source for much of the other data called for. If the procedural document has already been collected and appended to the work sheet, such data need not be repeated.

The "System or Group Preparing" block may contain the name of a single organization (or individual), or a category of persons, depending on the nature of the document. It might also contain the name of the automated system or program that produces the report. In this latter case, the working level group that specifies or is responsible for the production of the report should also be included. "Review Activity" should contain the name of the group or individual responsible for checking and approving the document before it is issued. "Distribution" will include those who receive the document for use or information and the number of copies they receive.

Spaces are also provided on the work sheet for recording various types of quantitative and reproduction information. All of this can be extremely useful in later determination of cost and benefit factors.

The bottom part of the form provides space for recording document contents, useful if that data is not already adequately described in an appended procedure, or obvious in an appended copy of the document. If data element information needs to be described, and detail beyond that indicated on the work sheet is called for, a supplemental Data Element Inventory (described later in this chapter) can be used and appended to this work sheet.

EXISTING-SYSTEM MANUAL FILE DESCRIPTION

Files are a critical system component and can significantly affect the dynamics of system operations. The method of retrieval used, for example, may significantly affect the response time of the system. Large files are expensive in terms of man-hours spent in their maintenance and in floor space occupied. Sometimes a change in retention policy can significantly affect system costs.

The Existing-System Manual File Description work sheet is illustrated in Figure 49. As with the previously described work sheet for document identification, this can be cross-referenced to the system flow. Space is also provided at the top of the work sheet for file identification data, including the file name, identifying number or code and location. The storage medium (such as "Five Drawer Letter-Size File Cabinet") is next specified.

EXISTING-SYSTEM MANUAL FILE DESCRIPTION **4.2**

Flow Sequence Ref. No. ________

File Name ________
Identification Number/Code ________
Location ________
Storage Media ________
Contents Description ________
____
____
____
File Purpose/Use ________
____
____
____
File Maintenance Responsibility ________
Users ________
____
Frequency of Access ________
Sequenced By ________
____
Volume/Size (Number of Records, etc.) ________
Rate of Change ________
Rate of Growth ________
Current Age ________
Data Retention/Purge Factors ________
____
Governing Procedure ________
File Accuracy ________
Unit Files:

Description	Number of Documents	Size

FIGURE 49: Existing-System Manual File Description Work Sheet

Spaces are next provided for noting a general description of file contents ("This file contains inspection records covering the past two years," for example), and describing the file's prime purpose or use ("This file is used for tracing complaints, and for developing a monthly report on complaints."). As to contents, the description can be brief for the analyst should be able to turn to the pertinent document identification work sheet for a detailed description of the file content.

The organizational group or individual responsible for the maintenance of the file should be noted next on the work sheet, followed by a listing of the prime users of the file. "Frequency of Access" refers to the number of times a day or week that a unit of the file is withdrawn or referred to. Unless this is some type of controlled file where access is documented, this will have to be an estimate. An average, but gross, figure should not be difficult to arrive at. Any restrictions on persons who can retrieve, on request, material from the file should also be noted, unless details are noted in the governing procedures.

Space is also provided in the work sheet for indicating the way the file is sequenced. This is followed by a space for recording file size. Size should be described in as many ways as are appropriate. Both the number of units contained in the files and the physical size (possibly in terms of the number of cabinets) are important.

In the Rate of Change space should be indicated the frequency with which the data or unit files are updated and the total amount of change in a given period. This might be stated in terms of transactions per day or a percentage per month. If inputs to a file are collected over a fixed period and all are filed at a scheduled time, this should also be noted.

Growth Rate, likewise, can be indicated in various terms but generally will be expressed as the net increase in unit files in a given period. Growth rate, age, and retention policy will be significant factors in forecasting needed system capabilities and may provide some interesting possibilities for cost savings.

The procedures which govern the maintenance of the file and its use should be referenced in the next space. For most, reference to a System Flow work sheet will be sufficient. If for any reason it is not, the pertinent details should be indicated on a supplement sheet.

Data accuracy will be more difficult to describe but some statement as to the reliability, validity, and timeliness of the data should be included. This will, of course, also relate to the accuracy of the input and the care taken in its preparation, and also to the frequency of update.

The "unit file" might also be described if more than one type of document is stored. Typically, a manual file might consist of conventional file cabinets in which the unit file is a manila file folder. Or it might be file boxes containing 3-by-5-inch file cards, or shelving containing loose-leaf binders. If there is a prescribed content to each unit file, that content should be described in the space provided. As for the content of the documents, themselves, once again the Existing-System Document Identification work sheets should be used.

EXISTING-SYSTEM DATA BASE DESCRIPTION WORK SHEET

In automated portions of a system, all of the processing programs operate on files or data retrieved from files. Because this retrieval operation (and the data retrieved) must be precisely defined in the programs, changes to a data base may result in considerable reprogramming and be more costly than changes to a manual file. Improving the match between operations performed and the structure of the file could greatly influence operating times and related costs. Therefore the detail with which a data base is described may be more critical than it is for a manual file. The work sheet provided for this purpose, Existing-System Data Base Description, is shown in Figure 50.

4.3

EXISTING-SYSTEM DATA BASE DESCRIPTION

(Automated Files)

Flow Seq. Ref. No.________

File (Data Base) Name________
Identification No. /Code________
Storage Media________
Contents Description________

Data Bank Purpose/Usage________

Frequency of Use________
Maintenance Responsibility________
Method of Updating________

Updating Frequency________
Updating Mode________
Rate of Change________
Rate of Growth________
Current Age________
Data Retention/Purge Factors________

Volume/Size (No. of Records, etc.)________
Description of Logical Record or Other Unit of Storage________

Structure and Sequencing________

Accuracy________
Methods of Retrieval________
Sub-Files________

FIGURE 50: Existing-System Data Base Description Work Sheet

On the other hand, if an automated system is well documented, there is more likelihood of the needed data already being recorded. If data called for by the work sheet has already been collected (in the Software Documentation, for instance), only a reference to its location is needed.

Much of the data called for on this work sheet is the same as that for manual files; however some additional data is needed to fully describe a data base. The method and mode of updating the data base should be noted, for instance. As to method, a reference to the programs that are used should be indicated. There should also be a statement as to whether single elements within a record can be changed, or whether only entire records can be replaced. The information noted here should also indicate whether or not the entire file must be passed at each update (as, for example, when inserting new records in a sequential file), and other similar pertinent information. These data are, of course, intimately related to those entered later under "Structure and Sequencing."

A description of the logical record or other unit of storage should be included on the work sheet. The contents of records identical to system documents can be detailed on document identification work sheets. If different, they can be described on the Data Element Inventory work sheets presented in the next section of this chapter.

The "Structure and Sequencing" block provides space for indicating whether the file is linear, sequential, or indexed sequential, or whether it has a more complex structure and is a randomly organized file, or uses one of the several forms of list organization. In the former case, the method for establishing the address of a particular record should be noted; in the latter case the type of pointers used should be described. Data base organization can obviously become quite complex and there are so many variations, some quite ingenious and also quite esoteric, that a reference to the actual documentation is probably the best entry that can be made in this block along with the briefest possible statement of the general category of organization used. The analyst should remember, also, that only information which will be of use later in designing system improvements or in determining costs or benefits need be recorded.

The same general comments can be applied to methods of retrieval. The most important item here is an indication of which data elements can be used as retrieval keys, and which are most often used.

Subordinate or derived files include those which contain some of the same data as the master file but in a different sequence or order. An index or inverted file is an example. They might also include a summary reflection on tape of a manual file. These should be indicated on the form and if sufficiently important, a separate file description prepared.

DATA ELEMENT INVENTORY WORK SHEET

As already indicated, the Data Element Inventory work sheet (Figure 51) is a supplementary form for both the Document Identification and the File Description work sheets. It is never necessary to make out more than one inventory for each document. If a file is simply a collection of formal documents, a copy of the document or an inventory may accompany the Document Identification work sheets and be appended to the File Description work sheet.

For an automated file, the machine version of the data rarely matches that of the written document. Therefore an inventory of data elements in each type of logical record in the file should be completed unless this information is included in the actual software documentation. This data will become of critical significance later should new programs be prepared for processing the same data.

In filling out the work sheet, the name of the data element should be the same as it appears on the printed form. For an automated system, the data element "name" will

4.4

DATA ELEMENT INVENTORY

Document or File Identification ______ Sheet ___ of ___

Data Element ______
Description ______
______ A/N ______
Configuration & Size ______
Other Considerations ______
Source ______
Use ______

Data Element ______
Description ______
______ A/N ______
Configuration & Size ______
Other Considerations ______
Source ______
Use ______

Data Element ______
Description ______
______ A/N ______
Configuration & Size ______
Other Considerations ______
Source ______
Use ______

Data Element ______
Description ______
______ A/N ______
Configuration & Size ______
Other Considerations ______
Source ______
Use ______

FIGURE 51: Data Element Inventory Work Sheet

generally be the code by which it is identified in the programs. The description column should define the content of the data element, elaborating on the name as necessary.

An indication of whether the data element is composed entirely of alphabetics or numerics, or is a combination of both, is of importance particularly in an automated system. This is the purpose of the A/N space.

If the data element has a fixed format, the required configuration should be indicated in the next column. A familiar example might be a date group which could be shown as "NN AAA NN" to indicate that it was always two numerics, three alphas, and two numerics (e.g., 05 MAR 82). If the element is of variable length up to a maximum, this can be indicated in whatever way is understandable to the analysts. Other conditions should also be noted such as running text of unlimited length, or sentences or phrases of an approximate average length. Some data elements may even consist of illustrations or drawings, or perhaps special symbols. The analyst must use his judgment in determining how far to go in describing this type of element.

"Source" can be important in estimating accuracy or in confirming system flow. Often reference to a procedure will be sufficient; sometimes reference to an organization will be the only possible entry.

In the "Use" space, a very brief statement of the use to which this data is put should be entered. Any data element for which there is no entry in this column should be examined very critically when analyzing the system for possible improvements.

DATA ELEMENT MATRIX 4.5

System/Sub-system ______

Document or File Ref. No. / Data Element								

FIGURE 52: Data Element Matrix

DATA ELEMENT MATRIX

As existing-system files are defined and documents accumulated and studied, the systems analyst should continually question the use of these data. Is a given report, for instance, merely received, then thrown away, or does it really serve some useful purpose? Are the number of copies received sufficient? Is the report worthless except for one or several data elements? Is the useful information on this report duplicated on another?

In examining and relating these factors the analyst may construct a number of various types of check-off lists and matrices. The Data Element Matrix work sheet, shown in Figure 52, serves both as an analytical tool and an inventory of the document and file work sheets. Its use is very simple but it can be very revealing. Data elements are listed by their categorical name in the left-hand column of the work sheet. Recorded across the top are the documents and files used in the particular system or subsystem being analyzed. These should be in the sequence they are prepared, if possible. A check mark is put in the grid square if a particular data element type is a part of the document whose name appears at the top of the column.

A number of interesting facts can surface when this matrix is completed. It may turn out that two separate reports are, superficially at least, identical. By that is meant they contain the same data elements. More often, they differ only by one or two elements and, through judicious planning, could be easily combined into a single document.

The matrix will generally also indicate, graphically, the way data elements are combined during the system flow, or how one data element will divide, paramecium-like, to form two subordinate elements during some operation.

8

Flowcharting the Existing System

The flowchart, a key part of the existing-system description covered in Chapter 6, is such a unique and special part of that description that it deserves to be treated as a separate entity. It is the graphic method for displaying a system's operation and sequence, and is sometimes referred to as the systems analyst's "shorthand." It can be used to pictorially describe both the existing system (as will be discussed in this chapter), and the proposed new or improved system (as will be discussed in Chapter 11).

There are basically two types of flowcharts. One is the program flowchart and the other the systems flowchart. The program flowchart, sometimes called "logic diagram," graphically portrays the data processing program logic. Program flowcharts are described in a later chapter. Systems flowcharts display the flow of information throughout all parts of a system, including the manual portions. Systems flowcharts can be of two types. One type is task-oriented, describing the flow of data in terms of the work being performed. The other is forms-oriented, following the forms through the functional structure of the system.

This chapter first describes the basic elements of flowcharting and describes some techniques used in their construction. Then it shows an example of using data recorded on the System Flow work sheet to develop a flowchart which, in turn, is attached to the Workbook's description of the existing system. Finally, there is a description of how forms, reports, and other documents are "keyed" to the existing-system flowchart.

The need for drawing flowcharts in a uniform manner depends on a number of different factors. If, for instance, an analyst is working alone drafting a flowchart of an existing system which he, himself, will subsequently use when designing improvements or a new system, then the need for uniformity is minimal. The only concern is for the analyst to adequately communicate with himself. In circumstances like these there was an analyst who, in his study of a court system, used unique little symbols representing court houses, judges, and so forth. For his own purposes, this proved quite adequate.

On the other hand, if there are other analysts on the same project, or if the job involves describing a system that someone else will program or, at some future date, modify, then there is a definite need for uniformity. A number of different groups, domestic and international, as well as several industrial firms, have issued guidelines on this subject. Some of these conflict in minor details. Generally, though, these "recommended practices," sometimes modified slightly in application, are accepted and understood. If the text that accompanies these symbols on a flowchart is explicit, then there should be no trouble understanding their use.

Basic Flowcharting Symbols

The basic flowcharting symbols most generally used are shown in Figure 53. These, when connected with flow lines, can depict virtually any system flow situation.

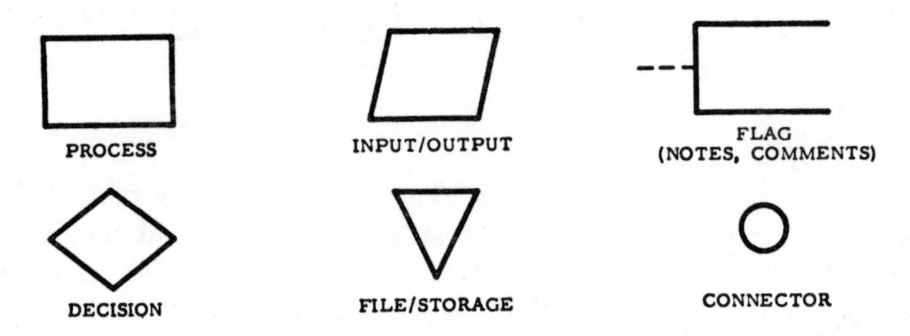

FIGURE 53: Basic Flowcharting Symbols

Process. Any manual or automatic operation or function. A process of performing some type of work. Example: "Multiply A × B," or "Compare Budgets to Expenditures."

Input/Output. Input material ready to be processed by the system, or output material produced by the system.

Flag. Used to append additional notes, comments, and descriptions that further clarify the call-out on the symbol to which the flag is attached.

Decision. A point in the flow where a condition is presented from which alternate actions (flow paths) can be taken. This can be a "yes" or "no" decision to a question, as an example, or it can be a switching type operation.

File/Storage. The filing or storage of material, regardless of form and regardless of storage medium.

Connector. A coded indicator that the flow continues somewhere else on the chart. The pick-up point is another connector symbol with the identical code.

In connecting these symbols with flow lines, normal flow is from left to right, or top to bottom, or a combination of these two directions. If the flow is other than this, arrowheads should be used to indicate flow direction.

A flowchart using all of these six basic symbols is shown in Figure 54. In this

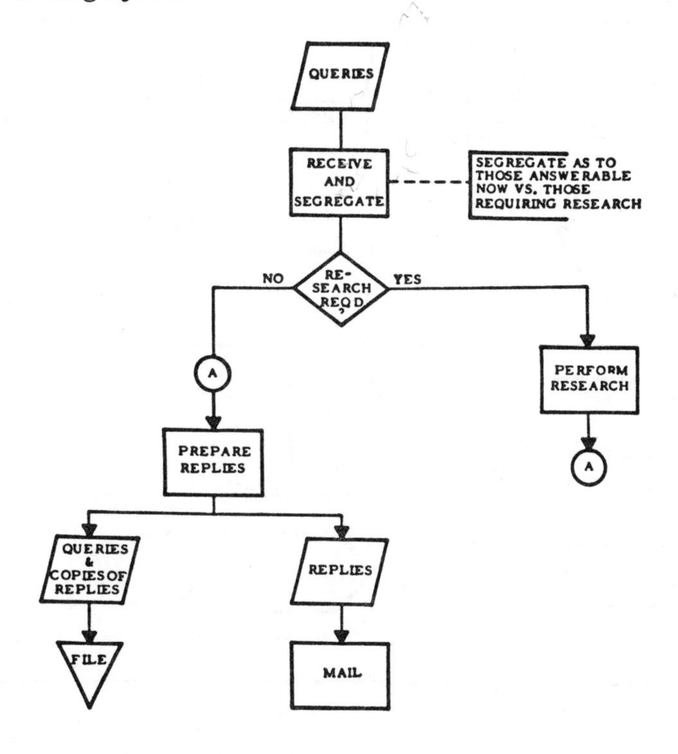

FIGURE 54: Example of Flowcharting Using Six Basic Symbols

example, input to the system consists of queries. In the first process they are received and segregated. The process of segregation is further clarified with a flag notation stating that the queries should be segregated as to those answerable now versus those requiring research before they can be answered.

Two processes follow. One is the performance of research which, once completed, joins the other flow (indicated by the connector "A"), where replies are prepared. Replies are mailed, while copies of the original queries, together with copies of the letters, are filed.

Relationship of Basic Symbols to Other Symbols

Figure 55 shows the basic flowcharting symbols as they relate to other symbols. These are used in various combinations for both program and system flowcharts. As can be seen by the way they are grouped in the illustration, these other symbols, at first glance somewhat complex in appearance, are really just more explicit definitions of the basic symbols. Thus, they can be used to more precisely display the flow of a system. Without their use, more words are needed in the basic symbols in order to describe the system operation. Definition of these additional symbols are as follows:

Manual Operation. Any process performed manually, such as "Visually Inspect."
Manual Input. Data entered into the system by push-button, keyboard, or switch settings.
Keying. Any type of keyboard operation, such as keypunch.

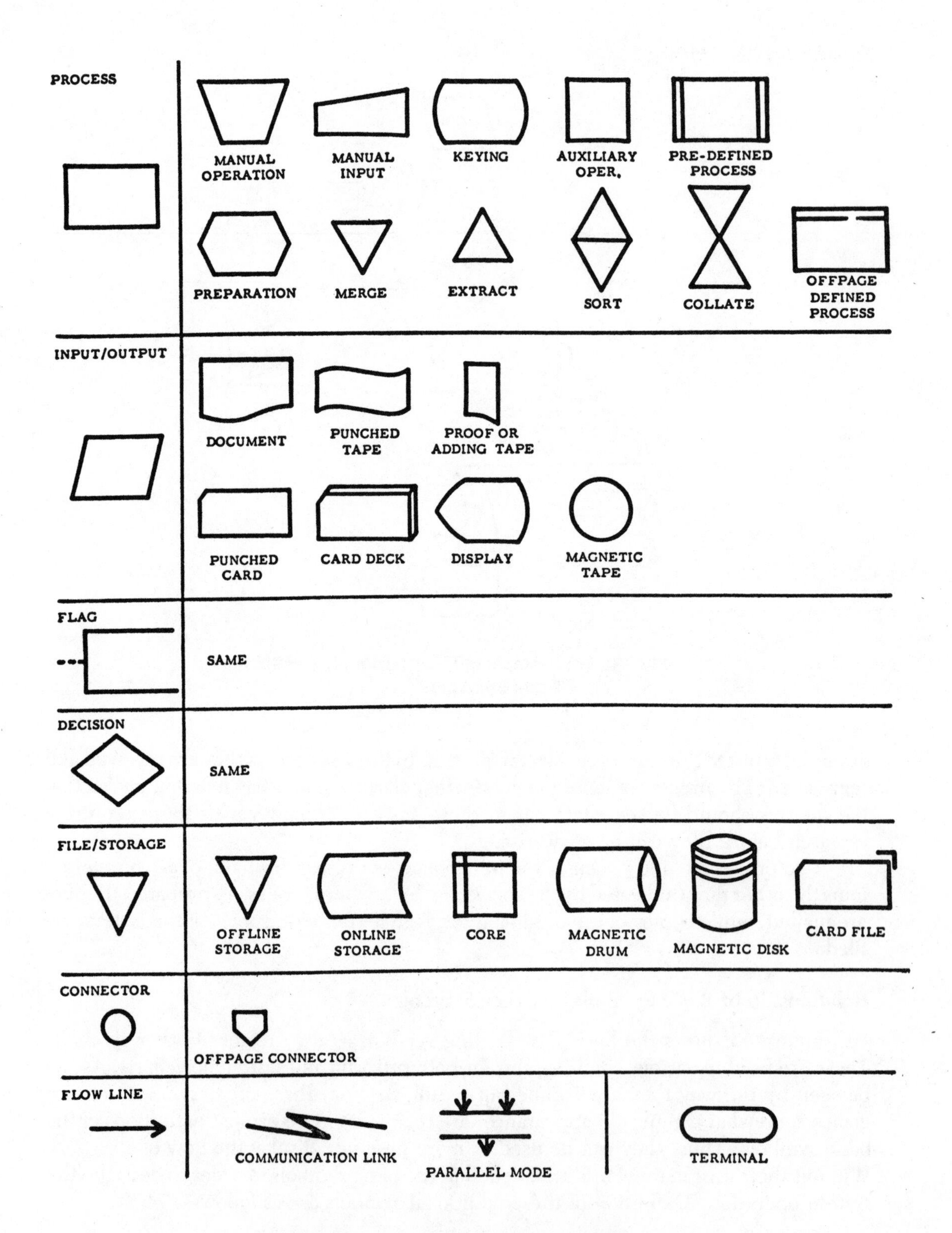

FIGURE 55: Relationship of Basic Symbols to Other Flowcharting Symbols

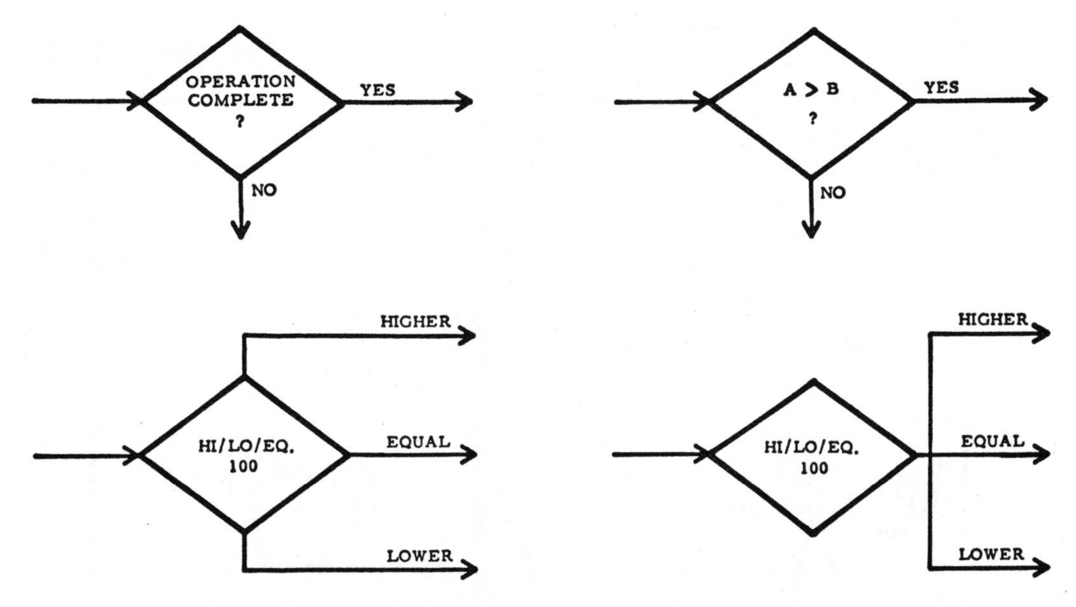

FIGURE 56: Various Methods of Using Decision Symbol

Auxiliary Operation. An offline operation performed on equipment other than the central processing unit.

Predefined Process. A sub-routine that is identified in the flow on the chart-at-hand, but is described elsewhere or specified in detail in another set of charts.

Preparation. An instruction to change or modify the basic program. An example would be to switch a segment on or off.

Merge. Combining two or more sets of data into one set.

Extract. Withdrawing one or more items from a set of items.

Sort. Arranging or ranking a set of items into sequenced order.

Collate. Comparing and combining several sets of items into two or more other sets.

Offpage Defined Process. A sub-routine that is identified in the flow on the chart-at-hand, but is described in detail somewhere else in the same set of charts.

Document. Any type of report, form, instruction, or other similar material that is used in the system.

Punched Tape. All varieties of punched or perforated tape used in the system.

Proof or Adding Tape. Adding machine tape or any other similar type of media.

Punched Card. All varieties of punched cards used either as input or resulting as an output from a system.

Card Deck. A deck of punched cards, although a single card symbol is often used to represent a deck.

Display. A display of information in some type of on-line mode such as CRT, printers, and plotting devices.

Magnetic Tape. A reel of tape, the tape possessing a magnetic surface on which data is stored.

Offline Storage. Any system or device for storage of material offline from the main system regardless of form.

Online Storage. Any type of online storage system such as magnetic tapes, drums, or discs.

Core. Magnetic core storage device.

Magnetic Drum. Magnetic drum storage device.

Magnetic Disk. Magnetic disk storage device.

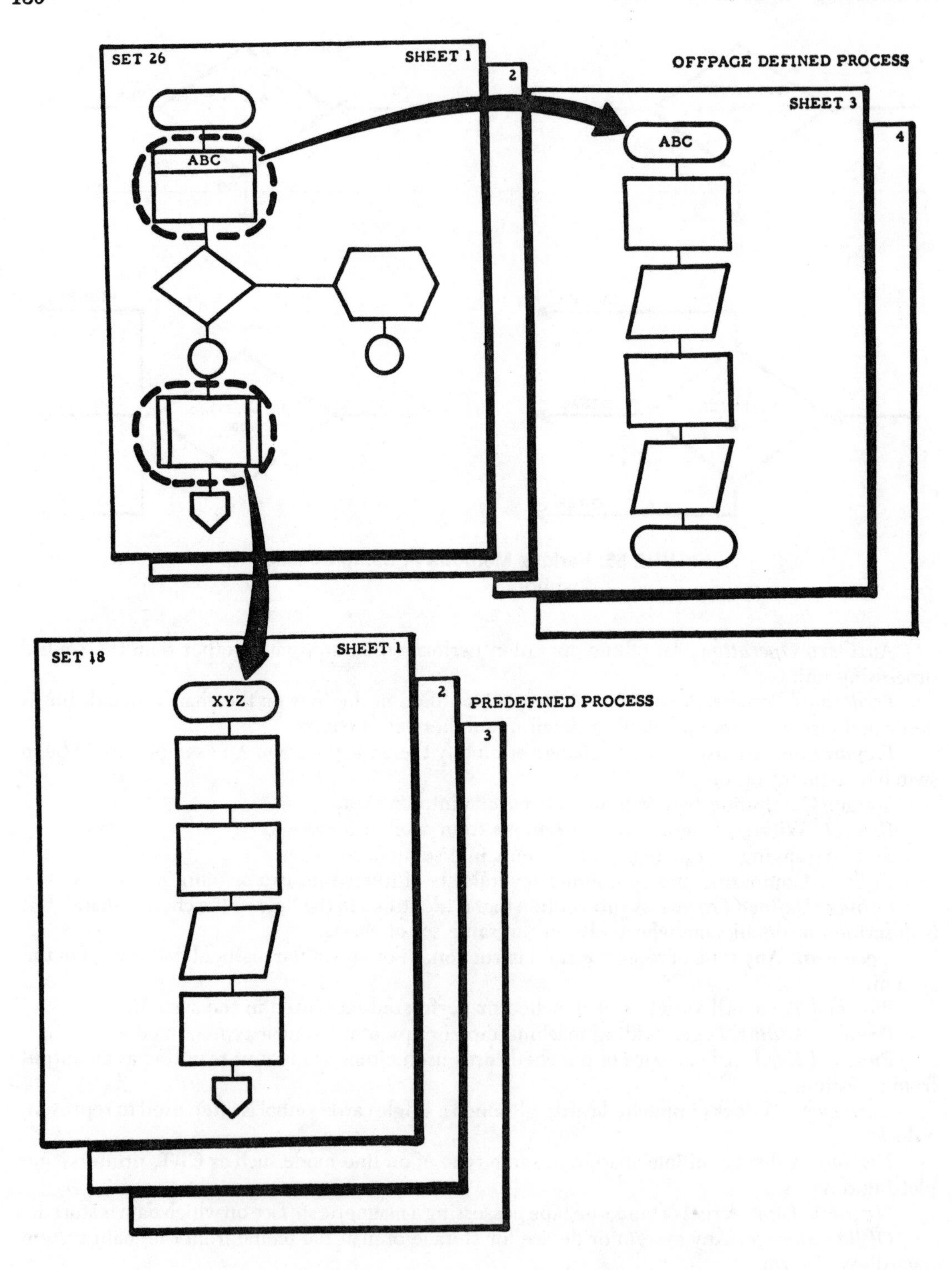

FIGURE 57: Predefined and Offpage Process

Card File. A punched card file cabinet.

Offpage Connector. A coded indicator that the flow continues on another page in the set of charts. The pick-up point is another identically coded offpage connector. (The original circle connector, under this expanded system, restricted for use as an "onpage" connector, only.)

Communication Link. Transmission of data by a telecommunication link. Arrowheads can be used to indicate directional and bidirectional flow.

Parallel Mode. The start or completion of two or more parallel operations.

Terminal. Designates a terminal point in a flowchart such as "start," "stop," "delay," or "interrupt." (Not used to indicate input/output device.)

Use of Symbols

It is sometimes useful to the understanding of these symbols to see examples of their use. Several examples of system flowcharts are displayed later in this chapter for that purpose. In addition, it can also be useful to amplify on the application of some of the more useful symbols.

Figure 56 amplifies on the ways that the decision symbol can be used. In the top left, for instance, the decision is whether or not the operation is complete at this point. Input can either be horizontal from the left, or vertical from the top. The example, top-right, shows a "yes" or "no" answer flowing from a decision point where a determination is to be made whether or not "A" is greater than "B."

The bottom two examples show a decision point that has the possibility of three exit paths. In these examples the decision is whether the input is higher, lower, or equal to 100. The method used at the lower right can also be used in showing more than three alternatives.

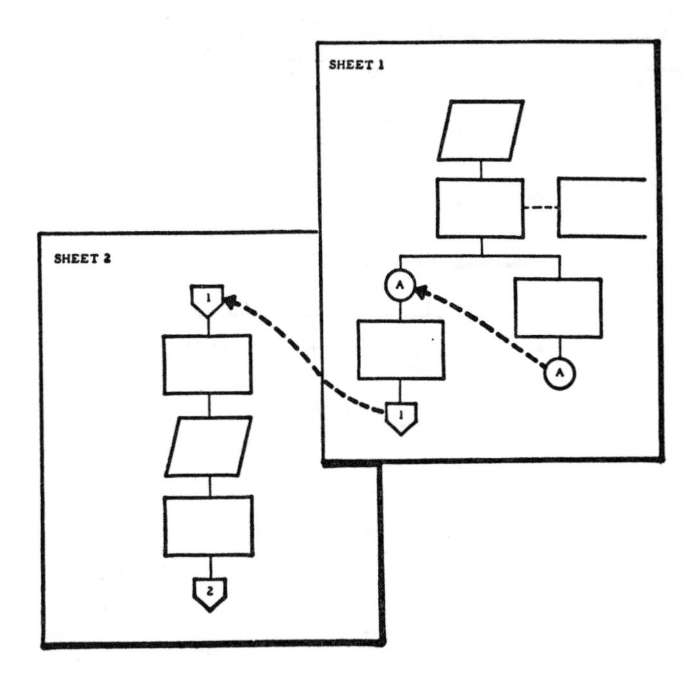

FIGURE 58: Use of Connectors

Figure 57 amplifies on the predefined process symbol and illustrates the use of another similar symbol which is the offpage defined process. In the predefined process, first shown on "Set 26" and identified as "XYZ" on the example, the detailed graphic definition of that sub-routine would be found on some other set of flowcharts (Set 18, on the example).

The offpage defined process, identified as "ABC" on the example, is similar to the predefined process, except its amplification will be found some other place on the same set of flowcharts ("Sheet 3 of 26" on the example).

The use of connector symbols is shown in Figure 58. The basic connector symbol (the circle) is used as an onpage connector, indicating the flow's continuation some other place on the same page of the chart. The offpage connector, identified with an appropriate code, indicates where the flow continues on some other page of the same set of flowcharts.

WORKSPACE AND AIDS

The development of the flowchart is usually done by the analyst in the privacy of his own work area. If he is "on location" and the client has not provided work space, this might even be his hotel room. In addition to this, if he is the type of analyst who prefers

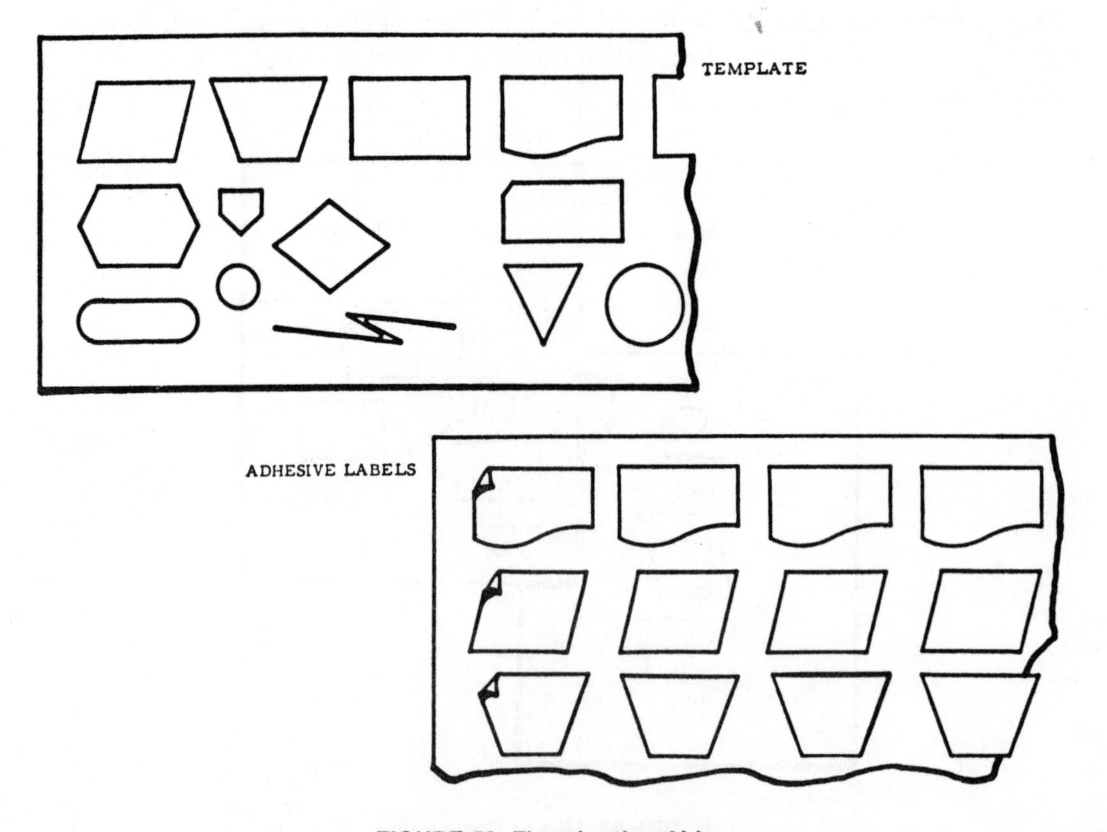

FIGURE 59: Flowcharting Aids

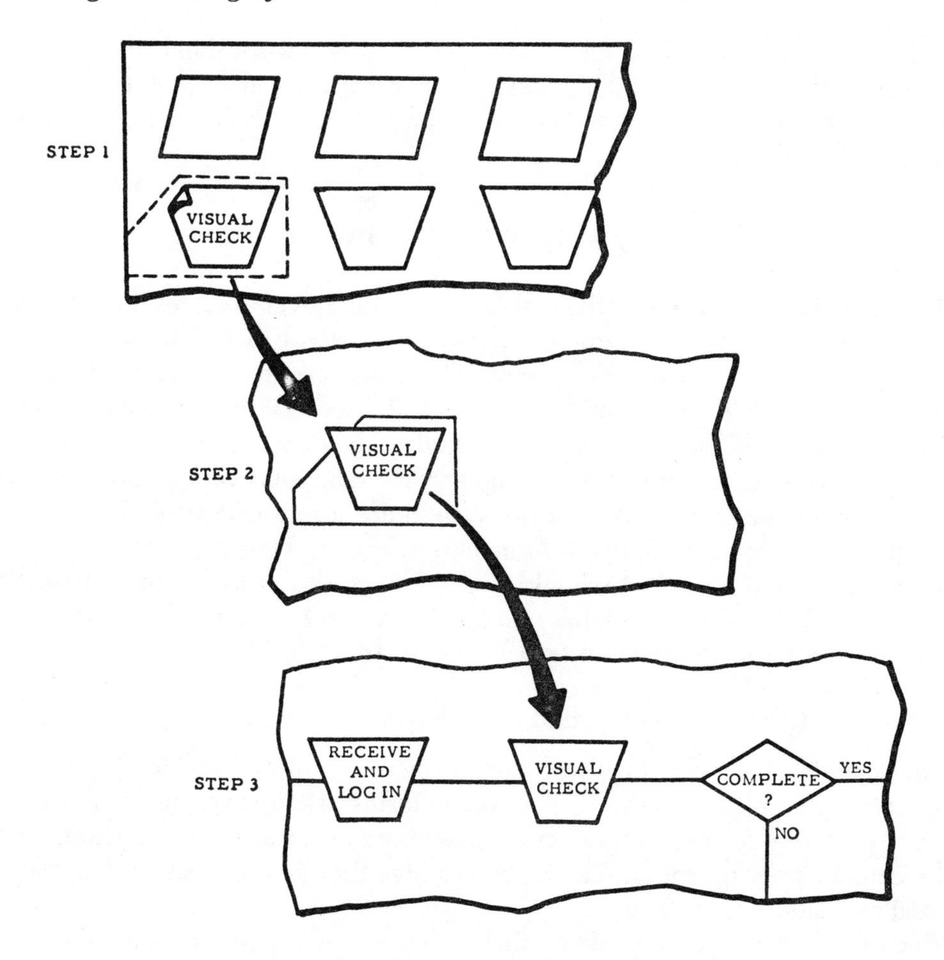

FIGURE 60: Use of Labels in Constructing Flow-charts

making big, continuous flowcharts rather than small individual sheets (using offpage connectors), then his work area might consist of his hotel room walls.

Flowcharting templates and adhesive labels, such as illustrated in Figure 59, are two of the more widely used aids employed by systems analysts in constructing flowcharts. The usual approach in using a template is to first sketch out a freehand rough draft of the general flow and processes before drafting a more formal chart. As the charting progresses it changes and is modified and symbols are frequently added, deleted, and moved. The more thoroughly the chart is planned and sketched in advance, the less grief there is for the analyst.

A method preferred by many analysts is the use of labels with adhesive backing. These can be transparent labels used on vellum if the analyst wishes to ultimately make blueline copies. Figure 60 illustrates a method for using flowcharting labels. In Step 1 the labels are shown with their adhesive backing. After the analyst writes his notation of the label, he cuts it out with its backing attached except with one corner of the label peeled back. In Step 2 he attaches the label with its backing to the work sheet, the label being held in place with one corner of the adhesive. After finishing placing all the labels and making all adjustments, he removes the backing and permanently attaches the label to his work sheet as shown in Step 3.

CHARTING THE FLOW

The prime input required by the systems analyst in developing his flowchart of the existing system is the System Flow work sheet described in the previous chapter. Supplementing this input are other elements of information derived from the Workbook. These consist of system environment factors, including a description of the existing data processing hardware (if there is any), and organizational data, especially useful if the analyst is constructing a forms-oriented flowchart. Examples of the existing forms, reports, and other documents used in the system are also a necessity. If the existing system software is documented, that information, too, is necessary input to the analyst.

The level of detail the analyst should resort to in constructing his chart depends on the job at hand. It should show detail to a level dictated by the needs of the analysis, rather than adhering to some theoretical "standard practice."

Task-Oriented Versus Forms-Oriented Flowcharts

Two examples of system flowcharts are shown in Figures 61 and 62. Figure 61 is a task-oriented systems flowchart and Figure 62 is a forms-oriented systems flowchart. The task-oriented flowchart shows a transaction involving the securing, for a customer, of an over-the-counter stock quotation. The forms-oriented flowchart displays the first steps of a proposal evaluation operation.

Which method to use depends on the analyst's preferences and the needs of the project. The only important point is that if there is more than one analyst, and their work must be merged, they must agree in advance on a single approach to flowcharting.

Using the Work Sheet

In Chapter 6 an example was presented of a dialogue between a systems analyst and an interviewee regarding the system flow relative to an operation that receives and stores goods. How the analyst recorded this information on the Workbook's System Flow work sheet was described and illustrated in Figure 41 of that chapter. How this material would look when translated into flowchart form is illustrated in this chapter in Figure 63.

In this hypothetical operation the receiving clerk receives the material which is accompanied by a packing slip. The clerk compares the received goods against his order file. If there is no order for these particular goods he returns them to the shipper. If the material and the order match, he then checks the material itself against the packing slip. If there is an error he adjusts his records and notifies Accounts Payable. If the material matches the packing slip, or if the records have been adjusted, he secures storage instructions for that particular type of material from his file and sends both the goods and the instructions to the stockroom. The packing slips are sent to Accounts Payable.

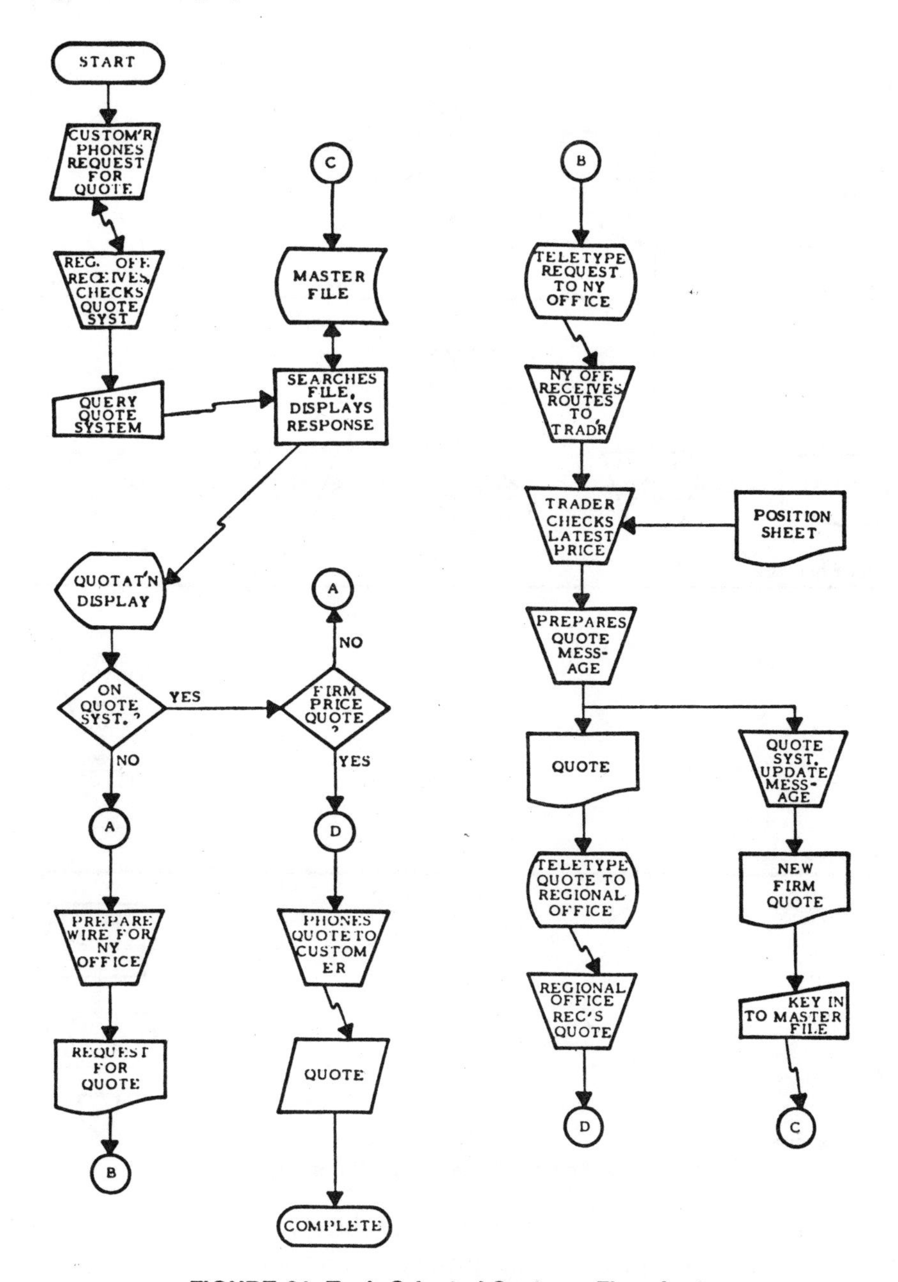

FIGURE 61: Task-Oriented Systems Flowchart

The stock clerk receives the goods. If the instructions for storing the goods are not with them, he returns the goods to the receiving clerk. If all is in order, he consults his stock code catalog and determines the proper stock number for this particular type of goods. He then identifies the goods with the code number.

He next updates his stock inventory system. This consists of a series of processes that have been defined in a procedure and identified with the code "S-21." The analyst

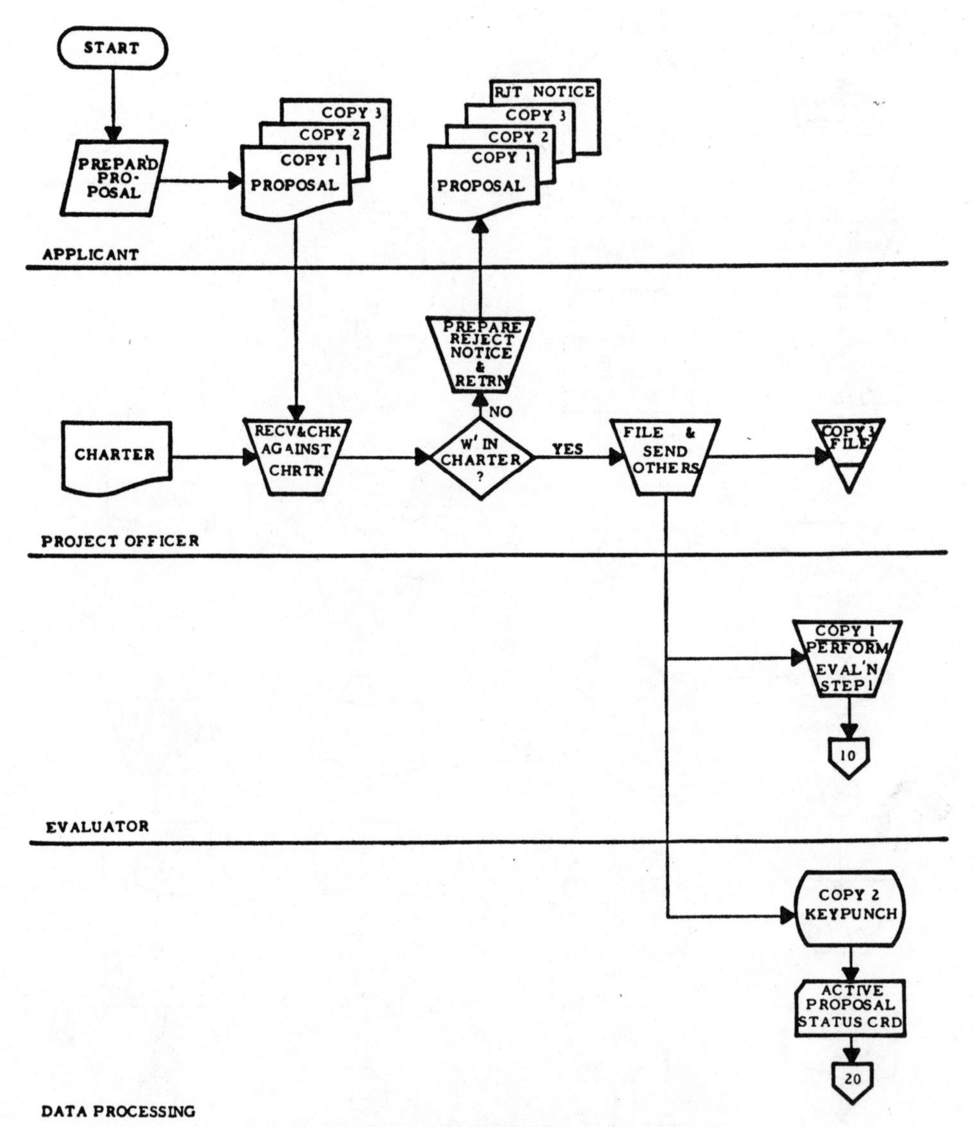

FIGURE 62: Forms-Oriented Systems Flowchart

intends to get a copy of this procedure later. After the stock inventory system has been updated the clerk stores the goods in their proper location and returns the instructions to the Receiving Department for filing in the storage instruction file.

The flowchart, once completed (as with the example just described), should be appended to the System Flow work sheets and inserted into the Existing-System Flow section of the Workbook.

CROSS-REFERENCING

As the systems analyst constructs his flowchart of the existing system, he should be referring, continually, to the forms, reports, and other documents used in the system.

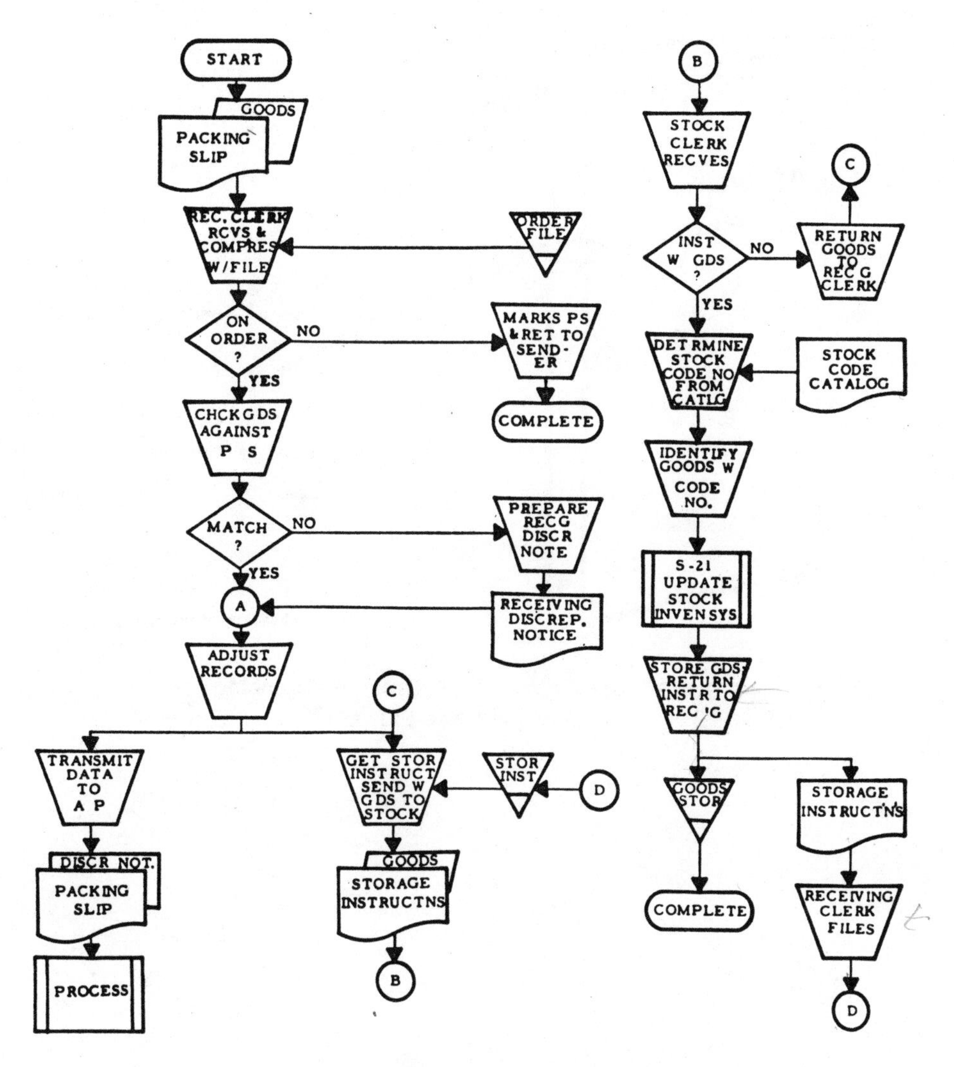

FIGURE 63: Example of Systems Flowchart Plotted from System Flow Work Sheet

Each of these should have a Document Identification work sheet attached to it. Once the documents have been noted on the chart, a reference number should be assigned. As shown in Figure 64, this common number should appear on the original document (and its Document Identification cover sheet as well, if so desired), and on the flowchart where that document is first indicated.

The documents, with their cover sheets, can be assembled in this sequence, and placed in the Existing-System Documents section of the Workbook for easy reference.

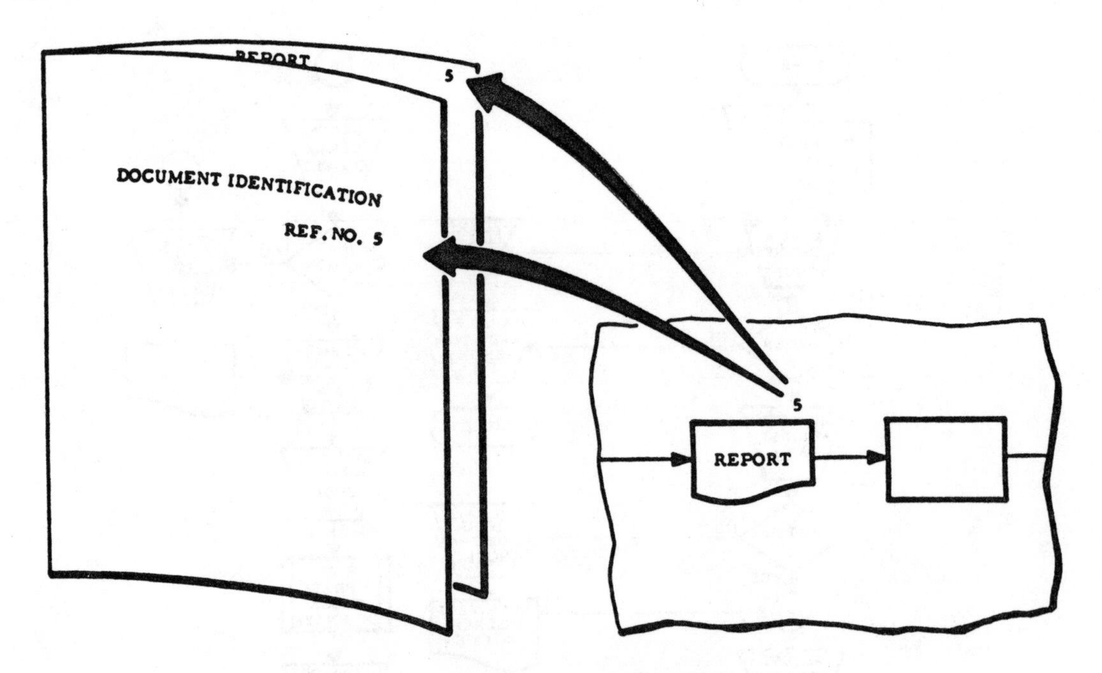

FIGURE 64: Example of Cross-Referencing Documents to Flowchart

9
Establishing a Cost/Benefit Baseline

Several years ago a system that had cost several million dollars to develop and implement, and about a half million dollars a year to operate was being updated to "third generation" equipment. The justification offered for this change was that it was "the thing to do technically."

Although a common argument, this is never sufficient reason to change a system. A change should be made only if it offers benefits that outweigh the costs of making the change. To properly approach a decision as to whether or not to make a change, these cost and benefit factors should be established and should constitute the only criteria for the decision.

Determining cost/benefit factors is a two-step proposition. First, the costs and benefits of the existing system must be established. This, then, becomes the "baseline" against which the estimated costs and benefits of the proposed new system are measured (the second step). This chapter deals with the establishment of the baseline.

COSTS, BENEFITS, AND VALUES

In any system, automated or not, the idea is that resources (in terms of capital and labor, translated into *costs*) are expended to achieve an objective (*benefit*) which has some worth (*value*). When the objective is a product or service which, in turn, is sold or directly affects sales, its value is easy to determine. When the objective (benefit of the expenditure) is a management report, or an accounting system, or a reservation system, its value is usually more difficult to measure.

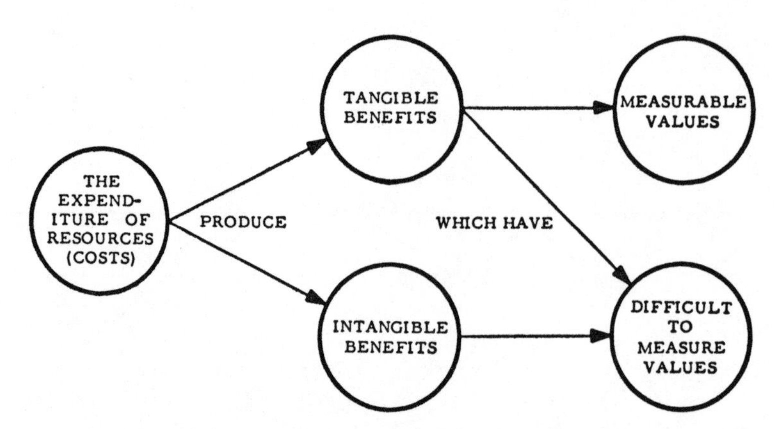

FIGURE 65: Cósts, Benefits, and Values

The relationship between costs, benefits, and values is illustrated in diagram form in Figure 65. It shows that the expenditure of resources can produce both tangible and intangible benefits. The values of intangible benefits ("goodwill," as an example) are always difficult to quantify. Tangible benefits can be easy or difficult to measure, or a combination of the two, depending on their nature.

In terms of the existing system (and later for the proposed system) the analyst must document operating costs, the products (benefits) derived from those expenditures, and some measurement of the value of these tangible and intangible benefits.

How all this works might be better understood by examining an actual analysis conducted for an airline terminal bus service. As it was being operated, passengers paid their fares to the driver upon entering the bus. The problems connected with this system included:

a. Congestion at the bus entrance (at both the airline and midtown terminals), frequently with the passengers standing, unsheltered, in the rain.
b. Buses frequently running behind schedule because of the initial congestion. Further, in speeding to make up for lost time the drivers were increasing the risk of accidents.
c. The need for drivers to carry excessive amounts of cash in order to make change for passengers who did not have the exact fare. (It was considered impractical to impose, or enforce, an "exact fare only" rule on passengers arriving from other cities.)
d. Frustration of the drivers when customers offered $20 bills as their "smallest change."
e. A rising rate of armed robbery of bus drivers, and the attendant hazard to passengers. This, in turn, caused an increase in insurance rates to the bus company.
f. Increased customer dissatisfaction because of all of the above.
g. Loss of productive driving time because of the necessity of having the bus drivers perform accounting tasks at the start and completion of each shift.

In documenting the costs and benefits of the existing system, the analyst was able to convert some of the factors, such as the robbery losses and the drivers' accounting time, into real dollar figures (measurable values). These were "tangibles" of the existing system against which the implementing and operating costs of any proposed improvement could be measured. Intangibles, such as customer dissatisfaction, were noted but not quantified. There was the probability that if the service improved more people might utilize it, but the analyst felt there was no realistic way to quantify this. (If subsequent to this it had been found that the cost of implementing and operating a proposed improved system would exceed the tangible benefits, the analyst would have probably found it necessary to try to justify his new approach by trying to measure this latter factor in some manner.)

The analyst's proposed new system offered a single, indoor, bullet-proof ticket booth at the midtown terminal to serve both incoming and outgoing passengers. The proposed system suggested the following advantages:

a. The ticket booth was less vulnerable to robbery than were the individual drivers.
b. $20 bills posed no problems, for a proper level of currency for change-making purposes could be safely maintained within the booth.
c. Automatic change-making devices, feasible with the use of the booth, offered the possibility of faster service to the customer.
d. If, despite faster service, there were waiting lines, these would at least be indoors where customers would be protected from the weather.
e. Bus drivers, no longer required to be cashiers, could be more productively scheduled driving buses. There would probably also be less frustration on the part of the drivers, as well.
f. With the removal of delays caused by ticket-purchase transactions at the bus door, schedules could be more properly maintained. There would be less reason for speed to make up for lost time, and less chance of accidents.

The analyst was able to demonstrate that the expenditure of a minimum amount of resources to implement and operate this new system would produce tangible benefits that possessed measurable dollar values (reduced losses, increased productivity, et cetera) superior to those of the existing system. Further, the new system would most likely produce some intangible benefits (increased customer satisfaction, reduced driver frustration), though these factors had values which were difficult to measure.

COMPARISON CATEGORIES

Data regarding the existing operation must be collected and grouped into three basic categories so that a meaningful comparison can later be made with the proposed system. These three categories are:

- Operating Costs/Benefits (for a stated unit of time)
- Life-Cycle Costs/Benefits
- Impact on Profit and Cash

Operating costs and benefits must be in some stated unit of time, like per hour, per day, per week, per month, or per year. To be more useful in the later comparisons with the proposed system, there should also be sub-categories of operating costs and benefits. In addition to "total system" the data could be arrayed to display these factors in terms of subsystems (segmented "runs" of the system, for instance), organizational units, and by units produced. This latter sub-category could be the costs required to produce a report, for example, or to process a charge card, or to secure a reservation.

The need for being able to later compare the existing and proposed system values over the "long haul" (life-cycle) requires the analyst to have information additional to operating costs. One type is the forecast information which should be already documented, at this point in the analysis, on the Plans and Trends work sheet. Also needed, at this point, is a definition of how the existing system's original acquisition/implementation costs are being amortized, and what those figures are. The determination of the *new* system's implementation costs is covered in Chapter 16.

If the system being studied is significant enough in scope that a change could possibly have a measurable impact on the company's profit and flow of cash, then there is still additional information to be documented by the analyst. This includes the company's most current financial information, such as the Balance Sheet, the Statement of Income, and the Overhead Statement. If the firm being studied is publicly held, this type of information is readily available. Later, the costs of the new system can be substituted for those of the existing system, with value examined in terms of profit dollars and cash flow.

MEASUREMENT APPROACHES

There are several approaches to determining the costs, benefits, and values of the existing system. One is to breakdown to a definable level all the tasks performed (as was done when planning the project). This is done by first extracting the operations portions of the existing-system flowchart. Once the breakdown has been completed, manpower, material, and equipment usage requirements and costs are determined. Another approach is to collect the required information at the organization level. There are advantages and disadvantages to both of these approaches, and the analyst must exercise good judgment in selecting that which is best suited for his particular project.

The analyst must also exercise good judgment as to the amount of detail to collect. In most cases it's beyond the realm of economic feasibility to conduct a precise, in-depth economic analysis. At best it's difficult to obtain good, solid, quantitative data. Under these circumstances it's usually better for the analyst to make a great number of calculated assumptions and estimates. This approach is acceptable if, after preliminary investigation, it becomes apparent that an uneconomical amount of time is required to collect the precise fact being sought.

THE WORK SHEETS

There are eight work sheets provided in the Workbook's Evaluation Criteria section for developing the existing system's cost/benefit baseline. The first two of these are "short form" type work sheets that can be used for documenting cost and benefit factors relative to simple, uncomplicated systems. The third is a more comprehensive form for itemizing costs in terms of the individual process steps of a system. The remaining five work sheets, used for more complicated systems and for establishing values of system benefits, are all titled "Existing-System Value Measurements" and are sub-titled as follows:

- Economy
- Efficiency, Productivity
- Quality, Usability
- Accuracy, Timeliness, Regulations
- Reliability, Adaptability

There is a direct relationship between these factors and the original system objectives set forth at the beginning of the project. The attempt here, then, is to try to quantify these factors in some manner, if that hasn't yet been done, so that proposed improvements can more easily be evaluated.

To attempt to pre-determine what specific type of value measurement information is important to any given project is a difficult, if not impossible, task. The work sheets provided, then, should be thought of in terms of guides. Although there will be some applications where the analyst could use them, as is, there will be many other cases where he must adapt and modify them to suit specific needs. As an example, in some instances the work sheets that are provided call for value judgments (poor, fair, good, very good, best). If the same factor can be more specifically quantified, then the analyst should certainly do so.

Once all the baseline information has been recorded on the work sheets, then this information is to be summarized on comparison-type work sheets, described later in Chapter 15.

The Existing-System Operating Costs Work Sheet

If the system being analyzed is simple and relatively uncomplicated, the analyst can use the Existing-System Operating Costs work sheet shown in Figure 66. The easiest situation would be where the total system operation takes place in one room, and everyone located in that room works full time only on that system, and if some type of equipment is required, that equipment is "captive" to the system, being used for no other purpose. Obviously there is no need to devise a very sophisticated cost-documentation system under circumstances like these.

In using the work sheet, the basic approach in determining costs is to establish a unit for costing, a unit cost, and a usage rate. The usage rate times the unit cost gives the total cost. The costing period can be anything appropriate to the task although the work sheet indicates total monthly costs. This is usually convenient to use and the period for which most usage rates are supplied. The exception may be in labor costs where the man-hour and the 40-hour week are the standard units.

The work sheet provides blocks for five general and one miscellaneous cost categories normally associated with any system. These are labor, material, supporting services, owned or leased equipment and software, and the miscellaneous category. No attempt has been made to break down these costs further on the work sheet since all systems will require different mixes of specific elements.

EXISTING SYSTEM OPERATING COSTS **6.1**

Labor Classification	No. Pers.	Manhours/Mo. ST	OT	T	Rate	Mo. Cost

Material/Supplies Description	Units Of Purchase	Units Used/Mo.	Unit Cost	Mo. Cost

Supporting Services Description	Units	Units Used/Mo	Rate	Mo. Cost

Equipment & Software Owned	Yr. Acquir, Orig Cost	Life	Amortz'd Cost	Maint. Cost	Mo. Cost

Equipment & Software Leased	Mo. Lease Cost	Maint. Fees	Other Costs	Mo. Cost

Miscellaneous Costs	Unit	Mo. Usage	Rate	Mo. Cost

Total Itemized Monthly Cost ______________

Indirect Costs (Rate ______) ______________

General & Administrative Costs (Rate ______) ______________

Total Monthly Costs ______________

FIGURE 66: Existing-System Operating Costs Work Sheet

Under the labor category the data needed will be the total man-hours used for each classification of manpower for which there are significantly different salary or wage rates. If overtime is required it can be handled in two ways. Overtime hours can be priced separately at the appropriate rate or they can be converted to equivalent straight-time hours and the converted hours plus the straight-time hours priced at the standard rate.

Materials and supplies directly chargeable to the system should next be recorded, followed by a list, and the cost, of supporting services. A column for indicating units has also been provided in this latter category although such services are often retained or

EXISTING SYSTEM BENEFITS DATA **6.2**

CATEGORY	DESCRIPTION	MEASUREMENT CRITERIA
ECONOMY/ PROFITABILITY		
TIMELINESS		
ACCURACY		
RELIABILITY		
SECURITY/SAFETY		
QUALITY		
FLEXIBILITY		
CAPACITY		
EFFICIENCY		
ACCEPTANCE/ USABILITY		
ADAPTABILITY		
OTHER		

FIGURE 67: Existing-System Benefits Data Work Sheet

charged on a flat monthly rate. The computation in these cases is simplified and only the last column need be filled in.

Equipment and Software have been combined, primarily because the computations used in arriving at monthly costs are roughly parallel. This category has also been divided into "owned" and "leased" classifications for convenience in computation.

Space is provided at the bottom of the form for summarizing costs and adding general and administrative costs, and indirect labor costs. Generally these are computed on the basis of set rates per hour or a percentage of the other total costs. The rates will vary among types of business and for different accounting methods and will depend on what specific costs are charged to these accounts. The customer's or in-house Accounting Department should be queried as to the proper factors to use.

The Existing-System Benefits Data Work Sheet

For small, uncomplicated projects, and as a supplement to the just-described work sheet for documenting operating costs, the Existing-System Benefits Data work sheet can be used. It is illustrated in Figure 67.

In using this work sheet it is necessary, of course, to document only those categories appropriate and significant to the system under study. Further, actual system characteristics (timeliness and economy, for instance) often overlap the rigid categorization imposed by a work sheet. Because of this, the analyst should concern himself less with classification, and more with an understandable description of the characteristic. Most important of all, the analyst should try to quantify some type of measurement of the described characteristic.

An example of how data might be recorded on the work sheet relative to a small study involving retail sales might be as follows:

- *Accuracy-Description.* "Inaccuracies in manually-summed sales slips."
- *Accuracy-Measurement Criteria.* "Of 560 average daily sales slips, 6% are incorrectly summed."

Other entries regarding the same problem might be recorded on the work sheet regarding the sales dollars lost because of these inaccuracies, the time required to make corrections, and other similar data.

The Existing-System Cost Breakdown by Process Step Work Sheet

A work sheet designed for itemizing costs in terms of a system's individual process steps is shown in Figure 68, Existing System Cost Breakdown by Process Step. A hypothetical case study is cited to illustrate this work sheet's application.

The system under study in this case involves the preparation and mailing of letters and price lists in response to queries received by mail. The flowchart in Figure 69 presents a graphic display of the process.

In this system, incoming mail is opened and sorted. Orders for products are transmitted to the appropriate department for order fulfillment. (This is a separate process which is not the subject of this study.) Queries for price information are delivered to the Sales Quotation Department where they are read, answers composed and typed, appropriate price lists secured, envelopes stuffed, sealed, stamped and mailed.

The problem with this current way of doing things is that the daily workload for the function is near the organization's capacity. This is illustrated in the Workload Versus Capacity Chart shown in Figure 70. The current capacity for answering price queries (utilizing five employees and five typewriters) is 170 answers per day. The workload is now at 160 per day. The forecast shows that this workload is expected to increase and will exceed the capacity within the next several months.

EXISTING SYSTEM COST BREAKDOWN BY PROCESS STEP

SYSTEM ____________________

Seq.	Process	✓*	Direct Costs Per				
			Payroll	Equipment	Supplies	Other	Total

*Place check mark at process step that is the pacing item. Describe limitation(s):

____

____

____

____

Subtotal	
Plus Fringe at % of Payroll	
Plus Indirect at % of Payroll	
Total	
Average No. of Units Processed Per	
Cost Per Unit	

FIGURE 68: Existing-System Cost Breakdown by Process Step Work Sheet

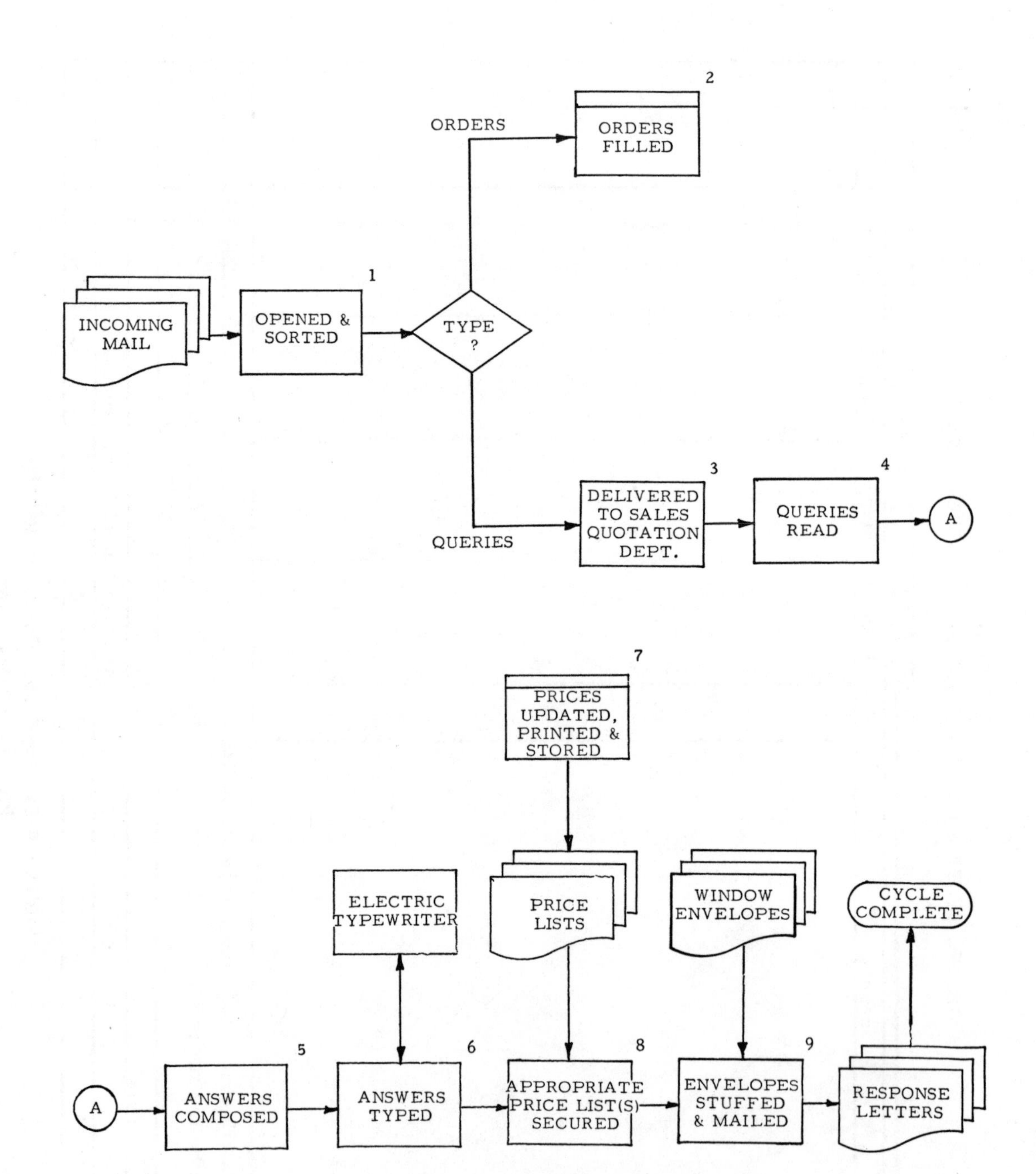

FIGURE 69: Flowchart of Existing Price Query Answering System

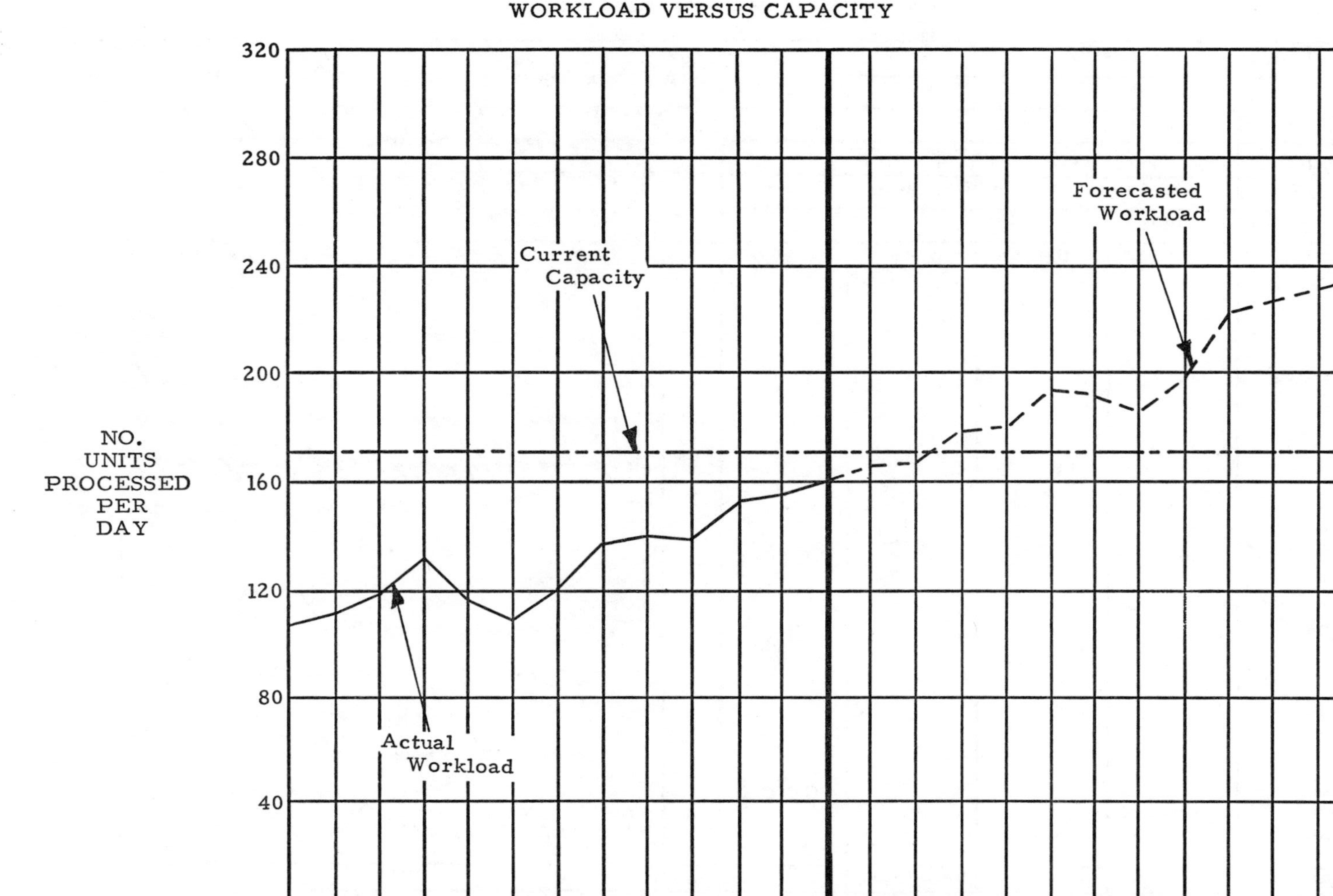

FIGURE 70: Workload Versus Capacity Chart

EXISTING-SYSTEM COST BREAKDOWN BY PROCESS STEP

SYSTEM Processing of Mailed-In Price Queries

Seq.	Process	✓*	Payroll	Equipment	Supplies	Other	Total
			Direct Costs Per Day				
1.	Letters opened & sorted. Orders or Queries?		$ 5.60				$ 5.60
2.	Orders fulfillment process (separate procedure)						
3.	Deliver queries to Sales Quotation Department.		2.80				2.80
4.	Queries read						
5.	Queries composed		186.00	5.77	7.20	10.40	209.37
6.	Answers typed						
7.	Price lists periodically updated, printed and stored (separate procedure.)					30.00	30.00
8.	Appropriate price lists secured.						
9.	Envelopes stuffed, sealed, stamped and mailed		11.20	1.20	3.20	15.60	31.20
						Subtotal	$ 278.97
						Plus Fringe at 12 % of Payroll	24.67
						Plus Indirect at 43 % of Payroll	88.41
						Total	$ 392.05
						Average No. of Units Processed Per	160
						Cost Per Unit	$ 2.45

(Steps 4–6 bracketed together; steps 8–9 bracketed together.)

* Place check mark at process step that is the pacing item. Describe limitation(s): Approximately 160 queries are answered each day by a staff of 5 people. Their capacity (5 persons, 5 typewriters) is approximately 170 per day.

FIGURE 71: Example Use of Existing-System Cost Breakdown by Process Work

In preparation for considering alternate design solutions to the pending problem, the analyst must attempt to document the cost of the existing system. This is done using the previously illustrated work sheet, now shown filled out in Figure 71, Example Use of Existing-System Cost Breakdown by Process Step work sheet.

The daily costs for this function, including factors for fringe benefits and indirect costs, are shown to total $392.05. Dividing this by the number of queries answered a day (160) provides a unit cost of $2.45. This becomes a useful number for later comparison with an improved system concept (see Chapter 15).

The Existing-System Value Measurements Work Sheets

The work sheets that would be used most often in documenting existing-system cost and benefit values are described in this section. These are the five Existing-System Value Measurement work sheets.

As previously mentioned, the work sheets described here might serve the needs of some systems analysis projects. More often than not, though, they should be used by the analyst as guides in developing work sheets more specifically appropriate to his particular project.

Economy. The first work sheet in the set, sub-titled "Economy," is shown in Figure 72. The first five lines of the work sheet, common with all other work sheets in the set, provide space for identification data. This includes the organization unit name, the task or system element whose characteristics are being documented on this particular work sheet, a number that cross-references this task or system element to the existing-system flowchart or System Flow work sheets, and the name or description of the primary output of this particular task or system element.

Next, some measure of the units of work performed in this task must be noted. If the task is to make credit checks, for instance, then it should be stated in terms like, "1200 credit checks per month." If the figure can be stated in terms of low, average, and peak levels, this would be even more useful.

The number of people required to perform the stated unit of work must next be noted. If they do not work full time on this activity, then the percent of time they do must be noted. This is followed by an indication of the skill levels used in performing this task, with the appropriate wage or salary rate or average indicated. If overtime hours are required on a regular basis this, too, should be noted together with an indication of the amount used.

If data processing equipment is used in this particular task or system element, it should be identified next on the work sheet. It is usually more meaningful to ascertain equipment usage factors on the basis of individual types of equipment, in which case if more than one type is needed for a particular task, separate work sheets should be used for each type.

The information to be recorded on the work sheets regarding equipment includes the estimated processing time per run or transaction, this being stated in terms of some unit of time. The cost rate is then noted, together with an identification of the method used for determining the rate. It may be, for instance, that this processing is done as a service for which there is a monthly bill. Or it may be that the existing data processing

EXISTING-SYSTEM VALUE MEASUREMENTS **8.3**

Economy

Organization Unit ______

Task (System Element) Performed ______

______

______ System Flow Ref. No. ______

Primary Output (s) ______

Units of Work Performed Per Stated Unit of Time ______

______

Number of People Required for Task ______

Percent of Time Devoted to Task ______

Skill Category	Wage/Salary Rate

Overtime Used Regularly? ______ How Much? ______

Data Processing Equipment Used for This Task (If More Than One Type, Use Supplemental Sheet) ______

______ No. of Units ______

Est. Processing Time: Per Run or Transaction ______

Per Day, Week, Month, or Year ______

Cost Rate Per Run or Transaction ______

Method of Determining Rate ______

Est. Cost Per Run or Transaction ______

Per Day, Week, Month, or Year ______

Significant Transmittals of Material ______ Cost Factor ______

Direct Chargeable Materials/Supplies ______ Cost Factor ______

Significant Travel/Transit ______ Cost Factor ______

Method & Amount of Charging Facility Usage ______

Other Resources Significant to this Task:

Type	No. Units	% Time Used on this Application	Cost Factor

FIGURE 72: Existing-System Value Measurements Work Sheet—Economy

system makes use of an accounting or usage control system that provides precise cost and usage statistics. Whatever the method, an estimated cost per run or transaction must be entered, designating its unit of time.

The next three lines on the work sheet are used for identifying and entering the costs involved with significant transmittals of materials, travel or transit of employees, and requisition of materials and supplies. If needed, supplemental sheets can be used to more precisely define these factors.

The bottom part of the "Economy" work sheet is used to identify types and costs of any other resources used for the task or system element identified at the top of the work sheet.

EXISTING-SYSTEM VALUE MEASUREMENTS — 8.4

Efficiency, Productivity

Organization Unit ______

Task (System Element) Performed ______

______

______ System Flow Ref. No. ______

Primary Output(s) ______

Graded Opinion of Management	Poor	Fair	Good	Very Gd	Best
Visibility Provided by Output	___	___	___	___	___

Name and Function of Grader ______

Comment Regarding Management Visibility ______

______

Degree of Standardization	None	Some	Total
Source Data	___	___	___
Input	___	___	___
Output	___	___	___

Equipment Utilization for this Task (As Identified on "Economy" Work Sheet):

______

This Equipment's Total Capacity (all units) ______

Capacity Available for this Application ______

Capacity Required for this Application (current) ______

Capacity Required in Terms of Anticipated Growth (per Plans and Trends Work Sheet) ______

Equipment Characteristics/Limitations ______

______

Cycle Complexity, This Task ______

______

______

______

List Redundant/Unnecessary Operations:

______

______

______

______

______

______

______

FIGURE 73: Existing-System Value Measurements Work Sheet—Efficiency, Productivity

Efficiency, Productivity. The second of five Existing-System Value Measurements work sheets, used for documenting factors relative to efficiency and productivity for each separate task or system element as with the first-described work sheet, is shown in Figure 73.

The first entry called for following the identification data is a "value judgment" regarding an opinion of management visibility provided by the previously specified output. The name and function of the person providing that opinion is an important factor that is also noted. There would be more weight, naturally, to the opinion of someone who must use the specified output, than someone who does not. The systems analyst

EXISTING-SYSTEM VALUE MEASUREMENTS **6.5**

Quality, Usability

Organization Unit ____________________

Task (System Element) Performed ____________________

____

____________________ System Flow Ref. No. __________

Primary Output(s) ____________________

Quality	Poor	Fair	Good	Very Gd	Best
Graded Legibility of Source Data	____	____	____	____	____
Graded Use of Abbreviations	____	____	____	____	____
Graded Neatness of Source Data	____	____	____	____	____
Graded Arrangement of Source Data	____	____	____	____	____

Name & Function of Grader ____________________

Comments Regarding Source Data Quality ____________________

____

Graded Legibility of Input	____	____	____	____	____
Graded Use of Abbreviations	____	____	____	____	____
Graded Neatness of Input	____	____	____	____	____
Graded Arrangement of Input	____	____	____	____	____

Name & Function of Grader ____________________

Comments Regarding Input Quality ____________________

____

Graded Legibility of Output	____	____	____	____	____
Graded Use of Abbreviations	____	____	____	____	____
Graded Neatness of Output	____	____	____	____	____
Graded Arrangement of Output	____	____	____	____	____

Name & Function of Grader ____________________

Comments Regarding Output Quality ____________________

____

Usability

Graded Usability of Output	____	____	____	____	____

Name & Function of Grader ____________________

Comment Regarding Usability of Output ____________________

____

Output Distribution Satisfactory? ____________________

Output Data Elements Satisfactory? ____________________

FIGURE 74: Existing-System Value Measurements Work Sheet—Quality, Usability

might also find it beneficial to secure value judgments from a cross-section of users. For this, separate work sheets should be used.

Another category of information called for on this work sheet is in regard to the standardization of the existing system's source data, input, and output. This is followed by several spaces for describing the total and available capacity of the equipment used relative to the task or system element identified at the top of the work sheet.

Equipment characteristics and limitations are to be described next, followed by a description of any complexities in the total processing cycle.

EXISTING-SYSTEM VALUE MEASUREMENTS 6.6

Accuracy, Timeliness, Regulations

Organization Unit ____________________

Task (System) Performed ____________________

____

____________________ System Flow Ref. No. ________

Primary Output (s) ____________________

Accuracy

Errors in Source Data ____________________

Errors in Keyboard/Transcription ____________________

Errors in Translation ____________________

Errors in Processing ____________________

Errors in Output ____________________

Validity of Computation or Algorithm ____________________

Average Number of Reprocessing Cycles ____________________

Significance of Errors ____________________

____

General Statement Regarding Accuracy ____________________

____

Timeliness

Elapsed Time Required to: Develop Data ________ Prepare Input ________

Process ________ Generate Output ________ Reproduce ________

Distribute ________ Total Cycle Time ________

Age of Data at Receipt by User: Min. ________ Max. ________

Days Between Issues ____________________

Elapsed Time Required to Respond to Special Query:

Minimum ________ Maximum ________

General Statement Regarding Timeliness ____________________

____

Regulations

Exceptions to Conformity to Regulations, Policies, Standards, Codes, etc.

(Derive from Requirements Statement) ____________________

____

____

____

____

____

FIGURE 75: Existing-System Value Measurements Work Sheet—Accuracy, Timeliness, Regulations

The bottom portion of the work sheet can be used by the analyst to record any redundant or unnecessary operations he has observed, or that have been pointed out to him.

Quality, Usability. Information regarding task or output quality and usability should be recorded on the work sheet shown in Figure 74.

Nearly all of the entries regarding quality are in the form of value judgments. These entries cover source data, input, and output, and call for judgments as to legibility, use of abbreviations, neatness, and the arrangement of the data. Specific problems can be

6.7

EXISTING-SYSTEM VALUE MEASUREMENTS

Relia ility, Adaptability

Organization Unit ______

Task (System Element) Performed ______

______

______ System Flow Ref. No. ______

Primary Output (s) ______

Reliability	Poor	Fair	Good	Very Gd	Best
Graded Reliability of Source Data	___	___	___	___	___

Name & Function of Grader ______

Comment Regarding Reliability of Source Data ______

______

Graded Reliability of Input	___	___	___	___	___

Name & Function of Grader ______

Comment Regarding Reliability of Input ______

______

Graded Reliability of Output	___	___	___	___	___

Name & Function of Grader ______

Comment Regarding Reliability of Output ______

______

______

Equipment Reliability:

Measured (__) or Estimated (__) Down Time in Relationship to This Task ______

______

Comments Regarding Equipment Reliability ______

______

______

Adaptability

Adaptability of Manual Portions of this Task to Handle Variable Workload ______

______

Adaptability of Existing Hardware for Handling Variable Workloads ______

______

Ease or Difficulty of Augmenting or Modifying Existing Software, Output Formats, etc. ______

______

______

FIGURE 76: Existing-System Value Measurements Work Sheet—Reliability, Adaptability

described in the "comments" sections. As previously stated, if the analyst is able to quantify these or other quality factors, he should do so.

Information regarding the usability of the output should be recorded on the bottom part of the form. This includes a value judgment and comments regarding output usability, as well as a notation regarding distribution and data elements.

Accuracy, Timeliness, Regulations. Figure 75 displays the work sheet to be used in recording value measurements in terms of accuracy, timeliness, and regulations. The first category, accuracy, calls for documenting errors occurring in various stages of the data being handled or processed. It is possible that this type of information is already being tracked and the statistics are readily available. If not, the analyst may have to resort to data sampling techniques. In any event, the information must be recorded in some type of measurable form, such as errors per week, or errors per document, or errors per transaction.

The validity of computations should be checked by the analyst, and the average number of reprocessing cycles due to inaccuracies and errors should be noted. Since there can be a wide range of significance regarding errors, a notation that highlights significant problems in this area should also be recorded. This should be followed by a general statement regarding accuracy.

In the timeliness category, elapsed times of the various steps in the task or system element cycle should be noted, followed by an indication of the minimum and maximum age of the information by the time it reaches the people who need to use it. Also called for is the time between issues, the elapsed time required to respond to a special query, and a general statement regarding timeliness.

In previous steps in the analysis of the system, regulations, policies, and other such material were examined in order to determine requirements. The last space on this work sheet is for recording in what areas the existing system deviates from those requirements.

Reliability, Adaptability. The final work sheet in the set of Existing-System Value Measurements work sheets, is used for documenting information regarding each task's or system element's reliability and adaptability. The work sheet is shown in Figure 76.

Source data, input, and output are to be graded in terms of reliability, together with an identification of the person who has provided the opinion, and a general comment regarding reliability. As with other value judgments, if actual quantitative measures are available or can be readily developed, they should be used in lieu of judgments.

Notations as to equipment reliability are called for next, including "down time" as related to this task.

The final part of the work sheet calls for definitions of the existing system's adaptability. This is in terms of the ability of the manual and hardware portions of the identified task to handle variable workloads, and the ease or difficulty of augmenting or modifying various parts of the existing system.

SUMMARIZING THE DATA

In developing the cost/benefits baseline of the existing system, all of the data collected is to be summarized on comparative-style work sheets which are described later in Chapter 15. The systems analyst may also find it useful to construct various graphic charts for the same comparative purposes.

10
The Value of the Workbook for Reviews

The Systems Analysis Workbook serves one of its key purposes when used for reviewing progress. Reviews can take two forms. One is in the nature of continuous reviews with the system users, the purpose of which is to make sure that the information they are contributing is complete and is being accurately documented. The other type of review is a more formal review with management. An appropriate time for the first formal review following the start of the project is after the existing system has been fully documented. At this point there are three categories of information that can be reviewed. First, of course, is a review of overall progress against the original agreed-to plan. Second is a review of the newly defined and expanded system goals so these can be added to or deleted from, as management sees fit. Third is a general review of all Workbook material accumulated to date from the standpoint of completeness and accuracy.

If the Workbook has been properly maintained, it is a much more useful tool for these reviews than a traditional "progress report," the preparation of which usually serves only to divert the analyst's energies away from the main thrust of his work. There is no need for the analyst to stop his work and feverishly prepare an impressive progress report for management. With the Workbook always up-to-date in terms of actual progress, the analyst is ready at any time, and at management's convenience, to review progress, ideas, and system goals.

The workbook approach to systems analysis, when thought of in terms of reporting to management, sometimes has some unexpected benefits. In one actual instance, a systems analysis was in progress at a military base for the purpose of developing an improved management reporting system. As so often happens in some branches of military service where rotation of management is an actively pursued policy, the officers who were the management personnel at this particular base were suddenly rotated. These were the officers who had authorized this particular systems analysis. When they were transferred to another base, a new set of officers moved in and took charge of the base being studied. Fortunately, because the Workbook had been used, there were no problems in quickly briefing the new base management on the status of the systems analysis project. They were immediately apprised of exactly what the goals of the analysis were, what had been accomplished to date, and what was being planned next. Furthermore, they were able to easily and conveniently interject some of their own ideas as to system goals. The analysis was able to continue productively, and to everyone's satisfaction.

TYPES OF REVIEWS

As previously stated, reviews fall into two separate categories. One comprises the continuous reviews with system users of the material as it is being developed. The other is the more formal review with management of all project material documented to date.

Continuous Reviews with System Users

One of the chief objections management has in dealing with analysts or "consultants" is that some of them use an "ivory tower" approach in their work. At the beginning of the project they appear briefly, ask a few questions, then disappear. They reappear at the end of the project with an innovative solution that may or may not have merit. Even those solutions with merit are often difficult to "sell" under these circumstances. The people who would have to live and work with the new system and could have possibly made worthwhile contributions to its design, resent and resist its implementation because they were never consulted. The Workbook approach to systems analysis avoids this pitfall, for its very makeup requires a close and continuous working relationship between the analyst and the people who are currently working with the system.

It is necessary, in this approach to conducting a systems analysis, that the developing information being recorded on the work sheets be repeatedly reviewed and corrected with the people who supplied the original data. It is an approach that emphasizes that the best foundation for constructing a new or improved system is the assurance that the background data and requirements have been fully and accurately documented.

It is really extremely difficult for an analyst to accurately secure all the information he needs in one sitting with the system user. The process of determining system flow, for instance, might go something like this. The analyst, using his System Flow work sheets, interviews the system user, recording the information as it is being discussed. He does this as fully and completely as possible. At the conclusion of the interview the analyst then constructs, from his work sheets, a preliminary system flowchart. As he constructs his flowchart he will invariably become aware of areas that are vague or that were not adequately covered in his notes. When he has completed his preliminary flowchart, vague areas and all, it will be about 90% complete and accurate.

The analyst then returns to the system user to review the flowchart in detail. Since flowcharts are not that difficult to understand, even if the system user has never seen one before, it usually takes only a brief orientation in fundamentals for him to understand it completely. Now, with the system flow laid out graphically the analyst carefully reviews each step with the system user, noting corrections and additions as they go. It is surprising how rapidly this can be done. At the conclusion of such a session, the marked-up preliminary flowchart is probably as accurate and complete as possible. Certainly it will be as accurate and complete as needed for the project.

Some of the material developed in the process of conducting a systems analysis can sometimes be of immediate usefulness to people operating the system. There might be, for instance, new employees in their department, and a copy of the flowchart might help in training these new people more efficiently and thoroughly than other methods they have had at their disposal. There should be no reason, unless their own management objects, for withholding "working" copies of this evolving material. If it is not yet complete, it only has to be so noted. This procedure also helps strengthen the relationship between the analyst and the system user.

Formal Review with Management

When a Workbook is used in a system analysis, management can at any time, and without serious disruption to work in progress, review progress. The most productive time for such a review, though, is when the current system and new-system requirements have been fully documented, and the cost-benefit baseline established. If the Workbook has been maintained in good order, all of the details will be readily available and there is no need for the analyst to go through elaborate preparations for such a review. All of the Workbook material to this point should be well organized and complete, and in understandable form.

The analyst should be prepared to review everything accomplished to date. The subjects that can be reviewed at this time are described in the following sections of this chapter.

REVIEW OF PROGRESS AGAINST PLAN

The entire project was started with the development of a project plan. In that plan the tasks to be performed were broken down, their interrelationships determined, and a schedule developed. Resource requirements were ascertained, costs and budgets established, and system objectives defined. All of these steps were accomplished regardless of the size or scope of the project, size and scope affecting only the amount of detail in the planning.

Planning is of no avail if there is no method for controlling to the plan. Again, the controls can be very elaborate if the magnitude of the project warrants it, or they can be very simple. Whatever the method of control, though, it is one of the purposes of a progress review with management to look at whatever control material has been developed and measure the progress that has been made against the plan.

The planning materials, themselves, such as the planning work sheets and the project network, are the basic tools used for analyzing project status. These materials are usually too detailed, however, for normal status review and reporting purposes. For that reason it is best to have developed, during the planning stage, simple graphic control charts. This is what was recommended in Chapter 2 so that now, at the first progress review with management, clear, graphic control charts exist and are ready for review.

The key word here is "graphic." The control charts must be designed so that the viewer can see the details of project status, yet quickly detect problems and other anomalies.

Control charts should all have certain features in common. First, they should show the plan with actual progress tracked against that plan. Unless a control chart has these two elements it has no validity, no purpose, and no value in project management.

Controls charts must also have provisions for readily and simply displaying changes in plans, "slippages," new forecasts, "recovery" plans and other deviations from plan. This information must be shown in such a manner that it doesn't "clutter" the chart with so much detail it no longer serves its graphic purpose.

Whatever graphic approach is used, the method of chart preparation should be one that does not require a great deal of time or artistic skill. It should not be a system that would pose any expensive or time-consuming reproduction problems.

Schedule Performance

The project network plan must be translated into some mode suitable for control purposes. This might be some form of PERT or PERT-Cost if it is a large project. For more modest projects the technique most often used is a milestone chart which utilizes key tasks extracted from the network. Such a milestone schedule, displaying tasks for a Phase I systems analysis project, is shown in Figure 77. This particular type of milestone chart has been designed to display the basic requirements of planned and actual accomplishment data. In addition, it provides for displaying major contingency information such as re-scheduling and slipped schedules.

An example of the same milestone chart updated for purposes of reporting progress is shown in Figure 78. Graphic clarity is achieved by shading all completed items to the left of the movable "time now" line. In the example, tasks 1111 and 1113 through 1151 were completed on schedule. Task 1112 is an item that slipped schedule one week but is now completed. Task 1311 is one that is not yet complete, but is on schedule.

As mentioned, shading on a milestone chart is intended to focus attention only on those items to the left of the "time now" line that remain incomplete, thus, behind schedule. Tasks 1152 and 1200 are examples of this. In the case of task 1152, it shows the anticipated completion date occurring in the 7th week rather than the originally scheduled 5th week. This anticipated slippage of two weeks is reflected in the shading, which ends two weeks to the left of the "time now" line.

Task 1200, Obtaining Interim Approval (Progress Review), has had to be re-scheduled as a result of the slippage of task 1152. The number "1" in the delta indicates that this is the first re-scheduling of this particular activity.

Other material could be added to the chart. For instance, dotted lines could be drawn between the activities in an attempt to show interrelationships. This, in effect,

PROJECT X MILESTONE SCHEDULE

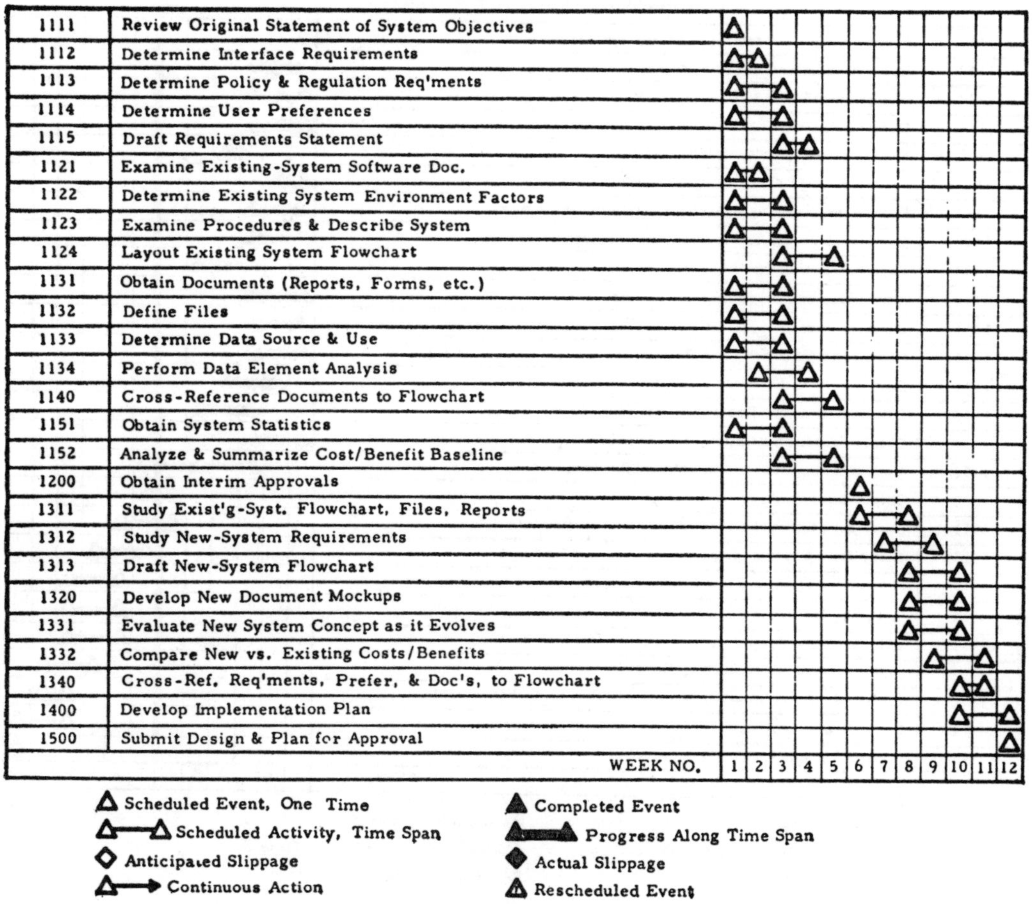

		1	2	3	4	5	6	7	8	9	10	11	12
1111	Review Original Statement of System Objectives	Δ											
1112	Determine Interface Requirements	Δ	Δ										
1113	Determine Policy & Regulation Req'ments	Δ		Δ									
1114	Determine User Preferences	Δ		Δ									
1115	Draft Requirements Statement			Δ	Δ								
1121	Examine Existing-System Software Doc.	Δ	Δ										
1122	Determine Existing System Environment Factors	Δ		Δ									
1123	Examine Procedures & Describe System	Δ		Δ									
1124	Layout Existing System Flowchart			Δ		Δ							
1131	Obtain Documents (Reports, Forms, etc.)	Δ		Δ									
1132	Define Files	Δ		Δ									
1133	Determine Data Source & Use	Δ		Δ									
1134	Perform Data Element Analysis		Δ		Δ								
1140	Cross-Reference Documents to Flowchart			Δ		Δ							
1151	Obtain System Statistics	Δ		Δ									
1152	Analyze & Summarize Cost/Benefit Baseline			Δ		Δ							
1200	Obtain Interim Approvals						Δ						
1311	Study Exist'g-Syst. Flowchart, Files, Reports						Δ		Δ				
1312	Study New-System Requirements							Δ		Δ			
1313	Draft New-System Flowchart								Δ		Δ		
1320	Develop New Document Mockups								Δ		Δ		
1331	Evaluate New System Concept as it Evolves								Δ		Δ		
1332	Compare New vs. Existing Costs/Benefits									Δ		Δ	
1340	Cross-Ref. Req'ments, Prefer, & Doc's, to Flowchart										Δ	Δ	
1400	Develop Implementation Plan										Δ		Δ
1500	Submit Design & Plan for Approval												Δ
	WEEK NO.	1	2	3	4	5	6	7	8	9	10	11	12

Δ Scheduled Event, One Time
Δ—Δ Scheduled Activity, Time Span
◇ Anticipated Slippage
Δ—► Continuous Action
▲ Completed Event
▲▬▲ Progress Along Time Span
◆ Actual Slippage
Δ Rescheduled Event

FIGURE 77: Example of Milestone Chart Used for Tracking Progress Against Plan

would be making a network out of the control chart and would only tend to clutter the graphics and defeat the control chart's basic purpose.

Cost/Budget Performance

Whether project costs and budgets are the subject of a review with client-management depends on the nature of the arrangement. If it is a consulting project being performed on the basis of a fixed price, then the current status of costs or budgets is of interest only to the analyst and the analyst's own management. If the contract is on some type of cost plus fee basis, or if this is an "in-house" type of project, then the cost or budget performance is of paramount importance to management.

A budget tracking chart such as shown in Figure 79 might be used in these latter cases. Depending on the needs of the project, and the number of people involved in the performance of the work, there might be a set of such budget tracking charts, the set

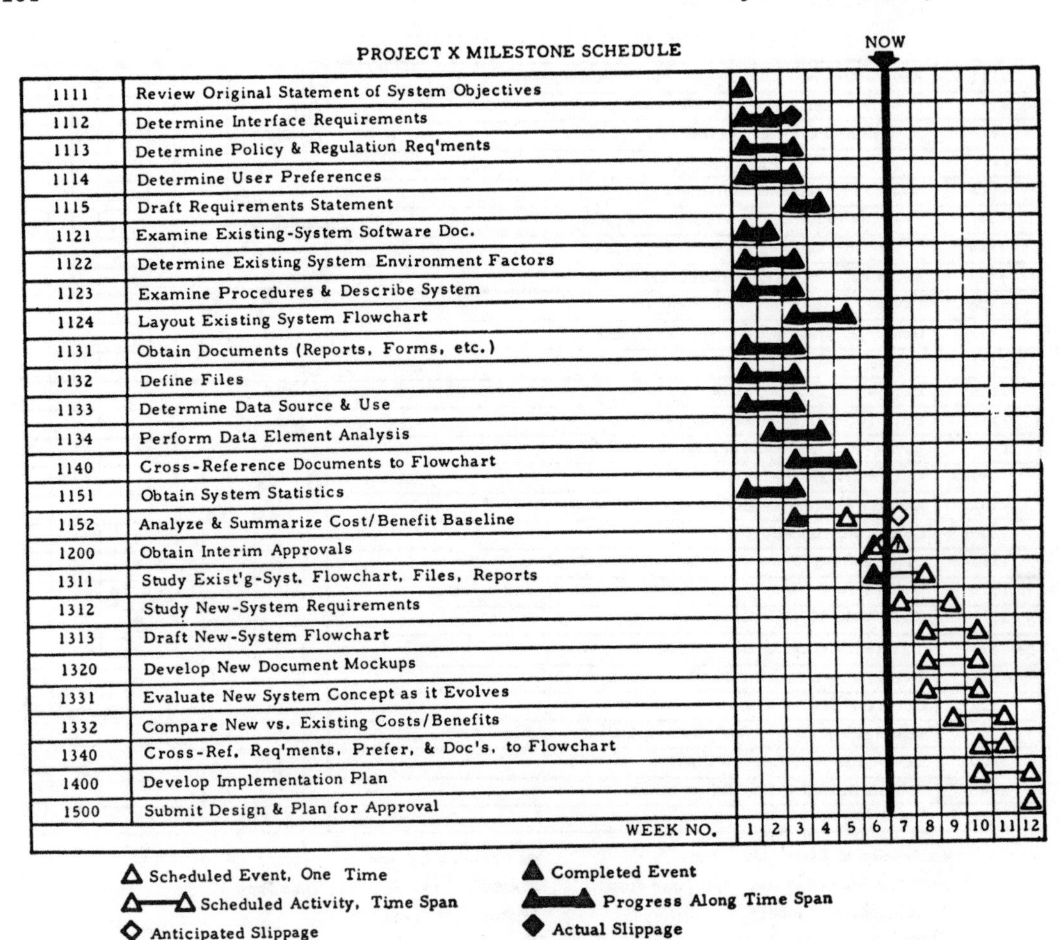

FIGURE 78: Example of Milestone Chart Updated for Progress Report

broken down in terms of who was performing the work, or by types of work being performed. The example, as it should be with all control charts, shows the actuals plotted against the plan. In this case, actual expenditures of the total project are displayed in relationship to the budget.

Relative to the format of the example chart (Figure 79), the actual statistics are shown in addition to the graphic portrayal of budget and actuals. In this case three lines of data are shown; budget, actuals, and the variance between the actual expenditures and the budget so that the viewer does not have to do any mental arithmetic. There is the option to add other lines of data, as well, such as budget and actual figures by the month rather than cumulative, as shown on the example. A line for entering a recovery plan or new forecast could also be included. Naturally, the more data that is included on the chart the more cluttered and less useful it becomes as a graphic aid.

Cost or budget performance is directly related to schedule performance, and this

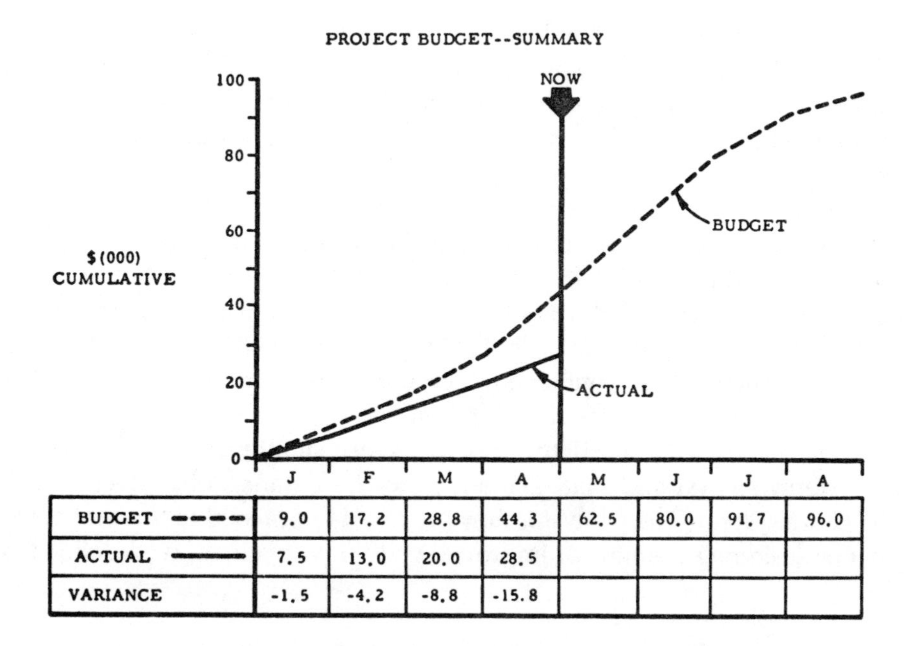

	J	F	M	A	M	J	J	A
BUDGET ----	9.0	17.2	28.8	44.3	62.5	80.0	91.7	96.0
ACTUAL ——	7.5	13.0	20.0	28.5				
VARIANCE	-1.5	-4.2	-8.8	-15.8				

FIGURE 79: Example of Budget Tracking Chart

relationship must be indicated. The project, for instance, might be "on budget" which when looked at separately may appear favorable. However, if only a portion of the tasks scheduled to be completed *have* been completed, then this would indicate, of course, an unfavorable condition.

Technical Performance

The opportunity for evaluating technical performance in the documentation part of a Phase I systems analysis project is extremely limited. It is primarily restricted to an examination of the system requirements which have been developed, and a review of the material generated to date for the purpose of determining its accuracy and completeness. Both of these subjects are covered in the following sections.

REVIEW OF OBJECTIVES AND REQUIREMENTS

The qualitative portion of the progress review with management involves examination of the original system objectives derived from the System Specification work sheet, compared with their refinement and expansion in the Statement of System Requirements.

Statement of System Requirements

In conducting the first part of the systems analysis the analyst examined the original Statement of System Objectives. He also determined, and recorded in his Workbook, interface requirements of the system. He has examined regulations and policies relative to their effect on the system. He has queried users as to their requirements and preferences. All of this he has summarized into a Statement of System Requirements. This statement must be transposed into an official new system "specification." This is done by reviewing the requirements statement with management and obtaining their concurrence.

Each specified requirement is examined in this review. There are undoubtedly particular requirements that management wants incorporated into the new or improved system. The statement is examined to make sure that these requirements are included. Subordinates may have specified a system feature that, from their standpoint, was a "requirement," a feature needed in order for them to perform their work. Management may or may not concur. That concurrence or rejection of the recommended system feature is indicated on the requirements statement.

At the conclusion of this particular review, there should be complete accord between the systems analyst and management as to the detailed objectives and the requirements of the system. The analyst is now prepared to proceed to the next step of the project, that of designing the new or improved system with a firm, well-defined goal in hand.

User Preference Checklist

In collecting system data, it will be remembered, the analyst not only collected information relative to system requirements (those items *necessary* in the design of a new system), but he also collected information as to user preferences (system features that would be *desirable* in a new or improved system). These preferences were then summarized into a checklist in the Workbook, a form that can easily and quickly be reviewed by management. This, once again, gives management the opportunity to concur or not with the desires of subordinates. Management can indicate those preferences they agree with and would like to see incorporated into the new or improved system. They can reject ideas if they wish. Or they can defer judgment until the following review when an evaluation on the basis of costs and benefit factors would have been completed.

System Features Susceptible to Improvements

During the course of the analysis, as ideas for system improvements were encountered, they were summarized on the System Features Susceptible to Improvements checklist in the Workbook. These were then "checked" as to those improvements that might be easily and quickly adopted now, if so desired, and those improvements that might be best suited for incorporating into the new or improved system design. It is at this progress review meeting that management has the opportunity of reviewing this particular material and indicating their preferences. There is no reason that management has to wait to get their money's worth from a systems analysis project. If there are improvements that can be made quickly and easily, management can initiate immediate action to have these things accomplished. Those ideas for improvements that, by their nature, are "long range" and would have to be thought of in terms of requiring design effort, can be added to either the Statement of System Requirements (if a necessary improvement) or the User Preference Checklist (if a desirable improvement).

These materials, the Statement of System Requirements and the User Preference Checklist, with all necessary and desirable features documented, become prime input to the design of the new or improved system.

REVIEW FOR ACCURACY AND COMPLETENESS

Every item in the Workbook is available for review by management at this progress review. The depth of detail that management might wish to pursue in this review depends on the circumstances of the project itself, and on the desires of management. It might be that they would be satisfied with only a cursory examination of this material, satisfied that their subordinates have examined and verified it in sufficient detail. On the other hand they may be extremely interested in every single element.

The more important Workbook items available for review, if a detailed review is advantageous, include the flowchart of the existing system, and all the forms and documents used in the system, keyed to the flowchart and organized in one section of the Workbook.

The current status of existing software can be examined, especially in the event that there's a conflict in this area. There is always the possibility that there is other conflicting or overlapping work being performed in the area being covered by the systems analysis. By reviewing what is occurring, such differences and conflicts can be resolved at this review.

The evaluation criteria that have been developed, translating the current system into terms of its current costs and benefits, in order to have a basis from which to compare proposed new improvements once they're designed, can also be reviewed. This permits management to fix in their minds a firm idea of the costs and problems that exist, and it prepares them to more decisively evaluate forthcoming proposed improvements.

System environment factors can also be examined from the standpoint of accuracy and completeness. If there are anomalies in this area these, too, should be covered.

Finally, the entire Workbook should be examined from the standpoint of completeness. If there are still missing pieces of essential data (which should be noticeable because of the blank spaces on the work sheets), management can suggest names of key personnel who might be able to help provide the missing data.

11
Using the Work Sheets to Develop the New System

When the documentation of the existing system has been completed, and the requirements and preferences for the new system have been defined and subsequently approved by management, the systems analyst has all the materials necessary for designing the new system or improving the existing system. The Workbook checklists and work sheets useful and necessary as input for this design effort include the system environment material, consisting of the description of the organization units and sub-units, historical data, and the plans and trends information needed to help the analyst in designing a system that will accommodate anticipated growth. There is also a description of the equipment that is currently used in the system, and the equipment that is on order, if any.

From a systems standpoint, there are the System Flow work sheets and the flowchart which describe the existing system. With this material are copies or descriptions of all the documents used in the system. The analyst also has detailed descriptions of existing automated and manual files. If the existing system is an automated system he has work sheets which describe it. If it has been documented, he probably has that software documentation, most likely including a program listing.

The systems analyst, in preparing to design the new or improved system, also has as input the Statement of System Requirements which lists every single necessary requirement that the new system must accommodate. In addition, he has in his Workbook the User Preference Checklist which lists the features that would be desirable in the new system. Finally, the systems analyst has the Evaluation Criteria work sheets which describe the costs and benefits of the existing system and indicate the design direction that must be taken in order to improve both of these factors.

Relative to automated systems, the extent to which the new design should be detailed is an essential consideration. A good rule is that the system should be designed only to that depth of detail which enables the programming and other implementation tasks to be accurately priced and scheduled. The management that authorized the systems analysis in the first place deserves the opportunity to examine the proposed system and readily make changes as they see fit, prior to a great deal of time and money being spent in such detailed tasks as coding and reproducing forms.

Other factors regarding the depth of design detail must also be considered, however. One factor is the skill of the programmers who would be assigned to the implementation phase of the project. Another factor closely related to this is in regards to whether or not any special or innovative programming concepts are to be used in the new system. If there are, most likely more detailed information should be communicated to the programmers. Work sheets for this purpose have been developed and are described later.

In developing a new system a number of activities take place, some of them in parallel with others. For purposes of describing these activities, however, they are grouped separately and described in the following sequence:

- Developing the New-System Flow
- Developing System Documents and Files
- Cross-Referencing

The development of equipment and facility requirements, if they apply, and training requirements, and the evaluation of cost and benefit factors, are also activities that are virtually simultaneous to the design of the system, but their importance necessitates their separate coverage in subsequent chapters.

DEVELOPING THE NEW SYSTEM FLOW

The activity around which all other design activities revolve concerns the development of the new system flow. Work sheets useful in various aspects of this activity are found in the New System Design section of the Workbook.

New-System Flow Chart

Chapter 8 dealt with the fundamentals of flowcharting and the method for developing a system flowchart of the existing system. As mentioned in that chapter, the ideas for developing a new system or improving the existing system usually occur to the systems analyst as he is analyzing and flowcharting the existing system. This being the case, one direct approach in the design effort is to begin the flowcharting task by marking up the flowchart of the existing system. This particular approach, naturally, depends upon the extent of the changes that are necessary.

Whether the systems analyst begins his flowcharting with a clean sheet of paper or by marking up the flowchart of the existing system, the important thing is that this is a systems flowchart. It can either be task-oriented or forms-oriented, according to the analyst's preference and the needs of the system being studied. The same symbols as described in Chapter 8 and the same flowcharting techniques are to be used. The idea is to develop a clear graphic description of the proposed system.

NEW-SYSTEM SYNOPSIS 7.1

SYSTEM NAME________

SYSTEM PURPOSE & SCOPE (Attach System Flowchart)________

GROUP RESPONSIBLE FOR OPERATION________

PROGRAM LANGUAGE TO BE USED________

MACHINERY SPECS________

MAJOR SYSTEM UNITS OR MODULES:

Unit or Module	Function	Source*

*Develop new, or salvage from existing system, or purchase/lease off-the-shelf

FIGURE 80: New-System Synopsis Work Sheet

System Synopsis Work Sheet

The first of the work sheets located in the New-System Design portion of the Workbook is the New-System Synopsis work sheet, exhibited in Figure 80. This work sheet, in effect, serves as a cover sheet to which the flowchart of the new system is attached.

This work sheet provides space for identifying the proposed system, briefly describing its overall purpose and scope. Space is also provided for identifying the organizational unit or sub-unit that will be responsible for the overall operation of the system. The program language to be used in the automated portions of the system is also specified on this sheet, as well as a summary of the equipment necessary to the system's operation

This might be merely a statement that says, "existing equipment will be utilized in the system," or "existing equipment plus (specify what's needed)," or "new equipment" (specify and attach supplemental pages, if necessary).

The bottom half of the form provides space for identifying the major system units, sub-units or modules that will be utilized in the automated portions of the system. Each unit or module is identified, its prime function briefly described, and the source of that module is specified. Regarding the source, this might be a unit or module of the system that must be developed new, or it could be an "off-the-shelf" proprietary type system or sub-system that can be purchased or leased. Possibly, it might be a unit or module that can be salvaged from the existing system. Relative to salvaging programs, in the past new generations of equipment were being designed at such a rapid pace, that most design efforts were involved with either converting manual systems to automated ones, or upgrading a system from one generation of equipment to the next. Salvaging parts of existing systems was seldom very practical. With the more sophisticated equipment that now exists, there's a longer life-cycle of equipment usage, thus enhancing the possibilities of salvaging parts of existing systems for use in new or improved systems. Salvaging, then, should be more prevalent in the future than it has been in the past.

If the new system being designed is a totally manual one, the bottom half of the work sheet would be used for identifying the sets of procedures, or the major elements of a single procedure, that would be required for the system. The "Source" column, in this case, would be used to indicate these either as existing sets or portions of procedures requiring little or no modifications, or as procedures that must be written new.

New-System Flow Work Sheet

The New-System Flow work sheet, illustrated in Figure 81, is also provided in the New-System Design section of the Workbook. It is identical, except in title, to the earlier-described System Flow work sheet used to document the flow of the existing system.

As with the chicken and the egg, there is the problem of whether to develop the new system first in narrative or graphic form. The choice is up to the analyst. Some prefer to develop the flowchart first, later transcribing it to narrative form on the New-System Flow work sheet. Because of the problems of handling complex graphics, other analysts prefer to do their "thinking" on the work sheet first, then translating it to flowchart form after all the "kinks" have been worked out.

A narrative version of the system flow is always a necessary adjunct to the design effort. The development of this narrative is, to many analysts, a nettlesome task. ("Why do I have to write out a description?" is a complaint often heard. "I'm a system designer, not Shakespeare.") Nevertheless, the narrative does serve several useful functions, one being to serve as the basic input from which operating procedures can be easily developed. Using the work sheet provided for this purpose accomplishes this and other purposes in a very useful manner. Another technique for deriving a narrative of the new-system flow, though less desirable than the work sheet, involves dictating this narrative by verbally describing the flowchart in sequence.

NEW-SYSTEM FLOW 7.2

System Element __________ Sheet ___ of ___

Prime Org. Unit __________ Governing Procedure __________

Seq.	Process	Con	Input/Output	Con	File/Store	Con	Decision	Con

FIGURE 81: New-System Flow Work Sheet

New-System Features Summary Checklist

As key system features are developed the systems analyst should record them on the New-System Features Summary Checklist, shown in Figure 82. As with the earlier described System Features Susceptible to Improvement Checklist, not all ideas for system improvements need wait for the completion and implementation of the new design. For that reason the analyst checkmarks each of these system features as to those that can be best incorporated in the new design, and those that can more easily be taken care of by a "quick fix." This special checklist, once created by the systems analyst in the process of his design effort, is reviewed by management at the same time that the proposed system design and implementation plan are being reviewed. If management concurs with the suggested quick fixes, action can be taken for their immediate adoption.

NEW-SYSTEM FEATURES SUMMARY **7.3**
Checklist

I.D. No.	System Feature	Check (✓) Implementation Opportunity	
		Quick Fix	New Design

FIGURE 82: New-System Features Summary Checklist

In the course of conducting a systems analysis, the analyst is invariably confronted with problem areas that are beyond the scope of the project at hand. Such ideas should be documented by the analyst (possibly through the use of the Project and System Specification work sheets located in the first section of the Workbook), and reviewed with management. Since this type of activity is most likely not defined as a project task and is therefore not a funded activity, the analyst should, in all fairness to the people paying the bill, devote only a minimum of time to this type of activity. Especially for consultants, anything more than a cursory step in this direction can be construed as the analyst's effort to merely try to expand the scope of his project and to extend his stay.

NEW-SYSTEM ENVIRONMENT FACTORS **7.4**

Changes in Organization Functions and/or Responsibilities:
(Check) Necessary____ or Desirable____
Specify____
____
____
____
____

Equipment Acquisition*
(Check) Necessary____ or Desirable____
Specify____
____
____
____
____

Facility Construction or Modification*
(Check) Necessary____ or Desirable____
Specify____
____
____
____
____

Personnel Training*
(Check) Necessary____ or Desirable____
Specify____
____
____
____
____

Other____
____
____

*Data summarized from detail work sheets

FIGURE 83: New-System Environment Factors Work Sheet

New-System Environment Factors Work Sheet

In designing the new system or improvements for the existing system, the systems analyst must consider various possible changes in the system's environment that are either necessary or desirable to the operation of the proposed system. For instance, changes in an organization's functions and responsibilities might be desirable, or even necessary, to the operation of an efficient new or improved system. This type of information should be recorded on the New-System Environment Factors work sheet, shown in Figure 83. A checkmark indicates whether the change is necessary to the new system's operation, or merely desirable.

As mentioned earlier in the chapter, the new or improved system might also require the acquisition of new equipment, construction or modification of facilities, special training of personnel in the use of the new procedures or in the use and operation of the system, and other similar factors. These are important enough as subjects to be discussed more fully in subsequent chapters in this book. The results of those detailed analyses, however, should be summarized on this particular work sheet, so that all such factors can be seen together in relationship to each other.

Manual Operations Procedure Requirement Work Sheet

The Manual Operations Procedure Requirement work sheet, shown in Figure 84, is also provided in the New-System Design section of the Workbook. A separate sheet should be filled out identifying each procedure that will have to be written or modified to accommodate the manual portions of the proposed system. The recommended procedure is identified by title and briefly described. The organizational units or sub-units that would be affected by this procedure are specified.

Since each step of the system flow is documented on the New-System Flow work sheets, those system elements apropos of this procedure are also referenced on this work sheet. As mentioned earlier, the System Flow work sheets are the most useful input to the writing of the manual procedures.

Space is also provided for indicating whether or not this replaces an existing procedure. If it does, that procedure is identified and a copy attached to the work sheet.

In dealing with manual systems, it is sometimes possible to design an improved system utilizing a simple change in procedure only. An illustration of this involved a repetitive, time-consuming system for reviewing and processing large production lots of plastic parts which had failed inspection. Each rejected lot represented a considerable amount of money, and the cost of scrapping or reworking the material was charged to whichever of several manufacturing departments was responsible for the defective condition.

For the purpose of determining responsibility and taking various types of actions, a number of departments had to be notified. Manufacturing management arbitrated the resulting disputes, Accounting established the dollar values, Production Control did the rescheduling, Engineering determined if design, machines, or procedures were at fault, Warehousing stored the materials, and Quality Control was responsible for the paper work and for re-inspecting, if needed. Under this system, a copy of the rejection report circulated through each of these departments, one department at a time. This took days or weeks. In some cases where it was to a department's advantage to delay matters, it even took months before disposition was decided.

The key bottleneck in the system involved obtaining the concurrence (and signatures) of the Production Control, Manufacturing, Engineering, and Quality Control department representatives, and it was precisely in these areas where the greatest time was lost.

The systems analyst spent a great deal of time poring over his flowcharts and forms trying to find methods to speed the flow of paper. The final solution came in the form of a

MANUAL OPERATIONS PROCEDURE REQUIREMENT 7.5

Procedure Title ____________________

Procedure Description ____________________

Organization Units/Sub-Units Affected ____________________

New/Improved System Flow Work Sheet Reference:

System Element(s) ____________________

Sequence Number(s) ____________________

Does This Replace Existing Procedure? If "Yes", Identify (and attach copy)

FIGURE 84: Manual Operations Procedure Requirement Work Sheet

directive and a simple procedure. All rejected lots were moved to a specially designated area. Every morning at 10 A.M. a representative of each of the departments met in that area, staying until each lot was disposed of and the rejection tickets signed. Processing time was cut drastically and so were warehousing costs. The representatives involved, who might have spent hours or days procrastinating when they could do this without leaving their desks, had much less patience with differences that required their presence. Disputes over responsibility were fewer and settled much more expeditiously.

DEVELOPING SYSTEM DOCUMENTS AND FILES

The activity of developing the new system documents and files must be done in conjunction with the other design activities. This involves the development of document mock-ups. It includes not only what are considered as normal input/output forms and reports, but can include catalogs, indices, letters, labels, notices, and any other type of document used in or produced by the system, automated or manual. Input for this design effort includes the work sheets that identify the existing system documents and files, the documents themselves, the Data Element Matrix, the requirements data, organizational descriptions, and the system flow work sheets and flowcharts. The new-system files that might be developed and described include both automated data bases, if any, and the significant manual files that would be used in the system.

New-System Input/Output Synopsis Work Sheet

The New-System Input/Output forms and reports are entered in synopsis form on the work sheet illustrated in Figure 85. On this particular summary type work sheet, the input and the output titles are specified. Next, some useful measure of document volume and size is specified. This might be the average number of pages in a report, or the number of records that are entered into the system, for instance, or some other similar indication of the amount of data involved. The final column is used for indicating the frequency with which the input and the output documents are processed.

New-System Document Identification Work Sheet

The work sheet provided in the New-System Design section of the Workbook to serve as a cover sheet for each New-System document mock-up, is shown in Figure 86. There is space provided for recording the document title and for indicating the system flow sequence number, or any other number for use in cross-referencing this work sheet and document to the flowchart.

The general purpose of the document is next specified, followed by a reference to the procedure, either existing or to be prepared, that will formalize the use of this particular document. Space is then provided for indicating the automated system that will generate this document, or the organization unit or sub-unit responsible for its preparation. The organization or person to be responsible for checking and approving the document before it is issued should be noted under "Reviewing Activity." Next, the document's proposed distribution and quantities involved in the distribution are also to be specified. The frequency of the document's preparation and distribution is also noted.

Entries are next made on the work sheet regarding an estimate of the annual quantity of this document, the media, the method of reproducing the blank form (and the size of the average reproduction order), the method of preparing the form, and the document size.

The bottom part of the form provides space for noting the details of the document's contents in terms of data element, size of that element, and the source of each data element. This type of data element information is more significant with systems where

7.6

NEW-SYSTEM INPUT/OUTPUT SYNOPSIS

Input	Vol/ Size	Freq.	Output	Vol/ Size	Freq.

FIGURE 85: New-System Input/Output Synopsis Work Sheet

there are fixed fields, of course, than it is with those types of systems where there are no limitations.

Data Base File Description Work Sheets

The backbone of a system is data. The often-used terminology for classifying data as it relates to a data processing system is as follows:

Item: A single piece of data, such as what might be represented by a single entry on a form (a person's name or social security number, for instance).

NEW-SYSTEM DOCUMENT IDENTIFICATION **7.7**

(Attach Mock-Up)

Flow Sequence or Ref. No.______

Document Title ______

Temporary Identification No. /Code ______

Purpose or Function ______

______

______

______

Formerly Covered by Procedure ______

System or Organization Who Will Prepare ______

Reviewing Activity ______

Suggested Distribution & No. of Copies ______

______

______

______

______

Frequency of Issue ______

Annual Quantity ______

Media ______

Method of Reproduction (Blank Form) ______ Quantity ______

Method of Preparation (Completed Document) ______

Size & No. of Sheets ______

Document Contents:

Data Element Title	Size	Source

FIGURE 86: New-System Document Identification Work Sheet

Record: A basic unit in a system, such as a source document (a purchase order or a time card, for instance).

File: A complete set of related records (sometimes called a data set).

Library: A collection of related files.

Data Bank (or Data Base): A collection of libraries.

Data records can be stored in file cabinets, in storage media such as punched cards, magnetic cards and tape, and paper tape, and in direct access automated storage devices of various types. Data files can be divided into several types:

Master File: A file of records constantly kept current, and which serves as the major source of information on a given subject.

NEW-SYSTEM DATA BASE DESCRIPTION 7.8

(Automated Files)

Flow Sequence or Ref. No.____________

File Name____________

Temporary Identification No. /Code____________

Proposed New File____________ Existing?____________

If Existing, Describe Required Changes & Modifications, If Any____________

Storage Media____________

General Description____________

File Purpose/Use____________

Anticipated Frequency of Use____________

Proposed File Maintenance Responsibility____________

Proposed Method of Updating____________

Mode of Updating____________

Initial Volume/Size (No. of Records, etc.)____________

Anticipated Rate of Change____________

Anticipated Rate of Growth____________

Proposed Retention/Purge Policy____________

To be Covered by Procedure____________

Description of Logical Record or Other Unit of Storage____________

Structure and Sequencing____________

Sub-Files____________

Methods of Retrieval____________

FIGURE 87: New-System Data Base Description Work Sheet

Transaction File: A file for containing inquiry or activity records which will be used for examining and/or updating a master file.

History File: A file of obsolete records and prior activities related to a master file.

Summary File: A file of data extracted from another file and condensed into summary form.

Trailer File: A file of data that augments information found in a primary file.

Two separate work sheets are provided in the Workbook for describing the files used in the system. One is the New-System Data Base Description work sheet illustrated in Figure 87.

As with the Document Identification work sheets, the flow sequence number, or

NEW-SYSTEM MANUAL FILE DESCRIPTION **7.9**

Flow Sequence or Ref. No.______

File Name ______

Temporary Identification No./Code ______

Proposed New File? ______ Existing File? ______

If Existing, Describe Required Changes & Modifications, If Any ______

Proposed Location ______

Storage Media ______

Contents Description ______

File Purpose/Use ______

Proposed File Maintenance Responsibility ______

Users ______

Anticipated Frequency of Access ______

Restrictions on Access ______

Sequenced by ______

Initial Volume/Size (No. of Records, etc.) ______

Anticipated Rate of Change ______

Anticipated Rate of Growth ______

Proposed Retention/Purge Policy ______

To Be Covered by Procedure: ______

Unit Files:

Description	No. Documents	Size

FIGURE 88: New-System Manual File Description Work Sheet

some other appropriate reference number, is indicated in the upper right-hand portion of the work sheet. The data base is identified in terms of name, type and temporary identification number. Whether this is an existing data base, or one to be newly created, is also noted. If it's an existing data base, required changes or modifications, if any, are briefly described.

Spaces are next provided for recording a general description of the data base, and for describing its purpose and use. The anticipated frequency of use should then be noted, followed by a designation of the organization or individual to be responsible for its maintenance.

The proposed method and mode of updating the data base should next be recorded with some indication of the initial size of the data base, along with estimates of the anticipated rates of change and growth. Coupled with this should be a notation of the proposed retention and purge policy.

A reference to the procedure that would cover the use and maintenance of the data base should next be noted. This might be a reference to a procedure requirement work sheet, previously described.

The description of the new-system data base is completed with definitions of the unit (or units) of storage, sub-files, the structure and sequencing of the file, and the methods to be employed in retrieving information.

The second of the work sheets is the New-System Manual File Description work sheet, shown in Figure 88. This particular work sheet is similar to that used for describing the data base, but has several added information elements, including the location of the file, and an identification of file users either in terms of individuals, or units or sub-units of the organization.

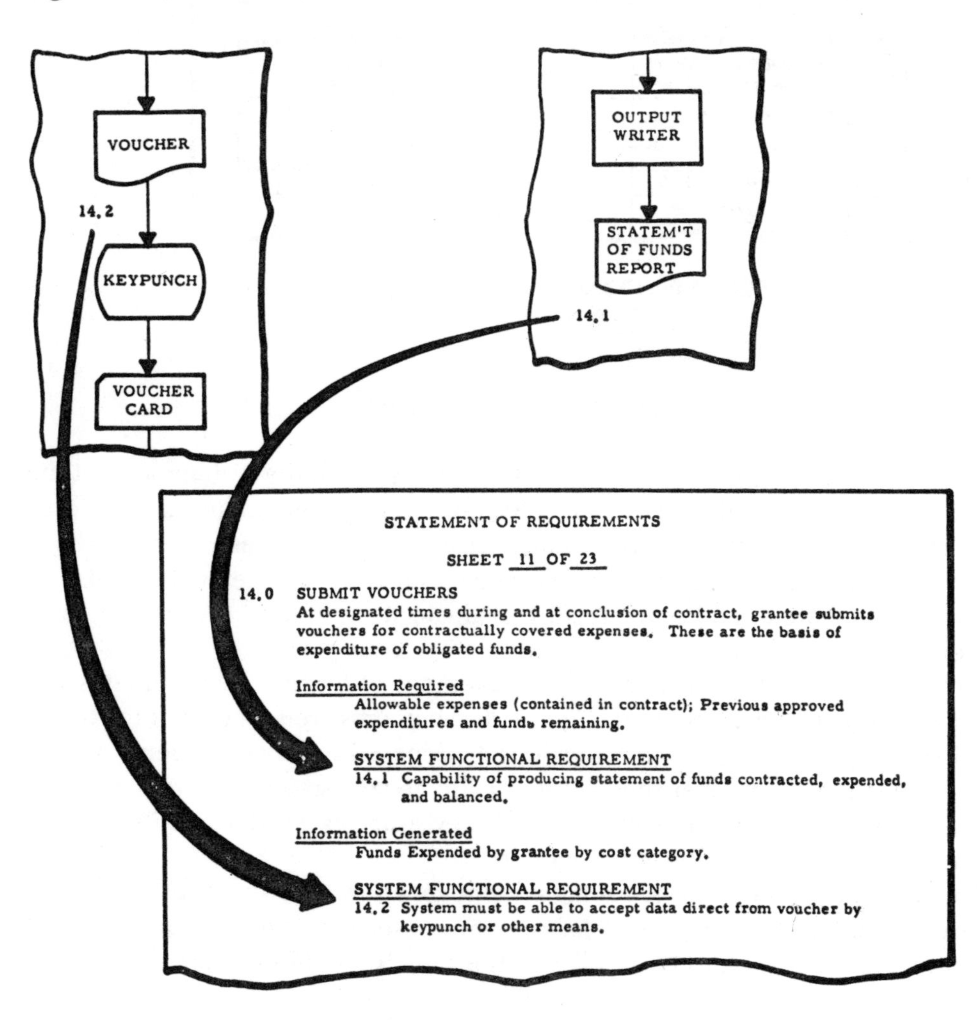

FIGURE 89: Cross-Referencing the Flowchart to Requirements

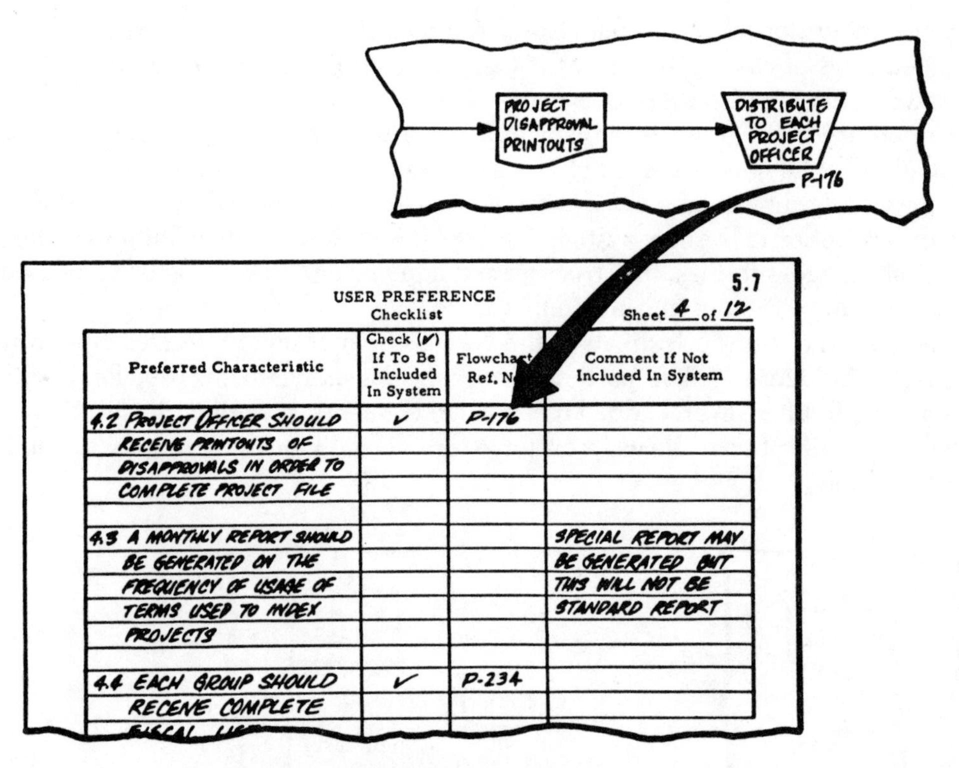

5.7

USER PREFERENCE
Checklist

Sheet 4 of 12

Preferred Characteristic	Check (✓) If To Be Included In System	Flowcha Ref. N	Comment If Not Included In System
4.2 PROJECT OFFICER SHOULD RECEIVE PRINTOUTS OF DISAPPROVALS IN ORDER TO COMPLETE PROJECT FILE	✓	P-176	
4.3 A MONTHLY REPORT SHOULD BE GENERATED ON THE FREQUENCY OF USAGE OF TERMS USED TO INDEX PROJECTS			SPECIAL REPORT MAY BE GENERATED BUT THIS WILL NOT BE STANDARD REPORT
4.4 EACH GROUP SHOULD RECEIVE COMPLETE	✓	P-234	

FIGURE 90: Cross-Referencing the Flowchart to Preferences

CROSS-REFERENCING

As the new or improved system flow is being developed, the systems analyst should be continually examining the Statement of System Requirements, the User Preferences Checklist, and the new-system documents, cross-referencing these items to the flowchart as it evolves. This type of cross-referencing enables management to more easily understand the system, and more readily evaluate its operation in terms of meeting requirements.

Figure 89 is an example of how the Statement of System Requirements can be referenced directly to the flowchart. Figure 90 shows cross-referencing of the flowchart to the User Preferences Checklist. Mock-ups of proposed system documents, with their Document Identification cover sheets, should also be cross-referenced to the flowchart. This can be done in the same manner that the existing-system documents were cross-referenced (described and illustrated earlier).

The particular numbering or coding scheme for any such cross-referencing is not a significant detail as long as it accomplishes its purpose. As suggested earlier, a simple, uniform coding system might consist of using the New-System Flow work sheet sequence numbers.

12
Developing the Conceptual Program Design

In dealing with automated systems, there are several factors that govern the depth of detail to which a system should be designed prior to the programming which takes place during implementation. As mentioned in the previous chapter, one governing rule is that the system must be designed to at least that level of detail where its implementation can be accurately priced. Going beyond this point imposes upon management the burden of paying for extra detail prior to "buying" the concept. The "pricing level" of detail automatically implies that this amount of detail is also sufficient input for the programming tasks.

In normal circumstances, the amount of detail already called for up to this point (and described in the previous chapter) should satisfy pricing/programming input requirements. If the programming is to embody some unique, innovative concepts though, additional information most likely needs to be communicated. Such concepts should be expected. With management's investment in the design effort, it's the analyst's responsibility to use ingenuity and creativity in developing the new system so that it offers a substantial improvement in operations.

This chapter first briefly describes program flowcharts, then presents several methods for expanding the definition of the system design.

PROGRAM FLOWCHARTS

System flowcharts, described earlier in this book, are used to display the overall operation of the system including both manual and automated functions. Program

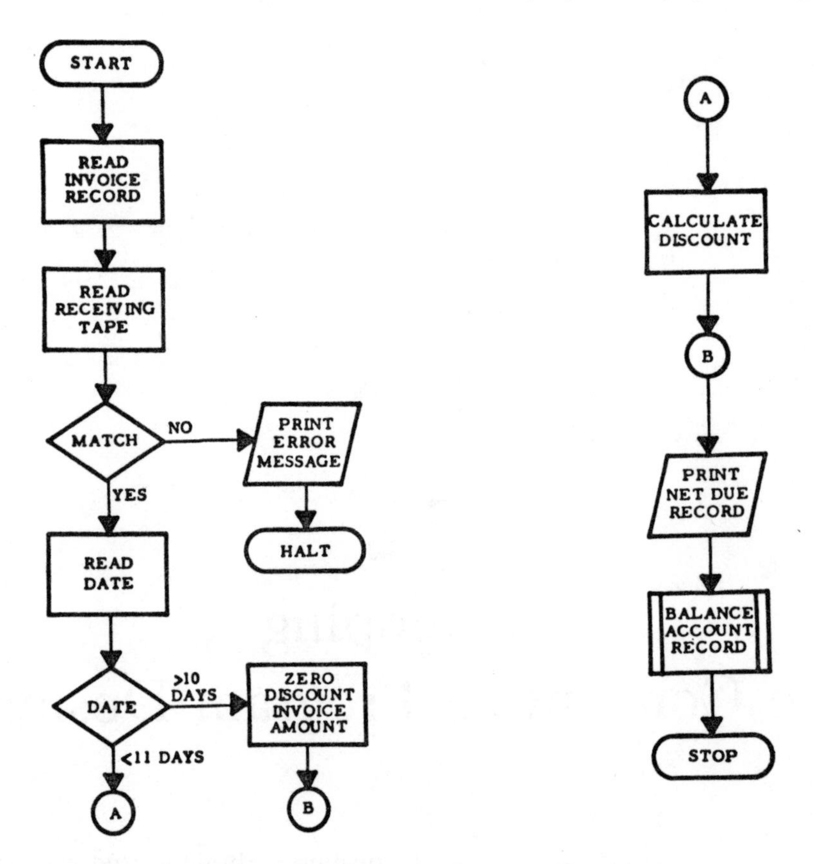

FIGURE 91: Example of a Program Flowchart

flowcharts, such as the one illustrated in Figure 91, are used to describe the individual steps of the data processing program.

Program flowcharts, sometimes called logic diagrams, can be developed to define several levels of detail. A "general" program flowchart is usually developed first. This provides an overview of the program's general logic. Next, more detailed flowcharts are developed, usually on the basis of separate logic diagrams for each computer run that has been designated. Finally, flowcharts are developed to show each discrete step in the operation. This is the definition needed in order to commence the coding function (translating the program logic into a symbolic code or "language" that can be "understood" by the data processing equipment).

If the need exists for the systems analyst to provide the programmer with a programming "concept," the analyst should develop a program flowchart that is at least to the "general" level of detail. The program flowcharting symbols and techniques of construction are similar to what was described in Chapter 8, Flowcharting the Existing System.

SPECIFYING PROGRAMMING DETAIL

Generally, the type of system design information that the systems analyst should prepare for the programmer consists of the system flowchart and description, the forms layouts, the documents and files descriptions, and the equipment configuration. Over and above this material, the analyst can also prepare a general program flowchart (Conceptual Program Design) and accompanying description, a list and brief description of the computer runs, and various other detailed program descriptions (in the form of flowcharts and/or decision tables) that amplify on any special or unique feature.

In order to better understand the programmer's needs, it is best to first briefly describe the programming task. The programmer, after receiving at least the minimum of the material specified in the foregoing paragraph, defines the system in greater detail through the use of flowcharts, decision tables, descriptions, and other means. Once the program logic has been described in sufficient detail, the coding function takes place. The material on the coding sheets is then keypunched onto cards and a printed listing is produced. The listing is checked for errors, corrected, and a new listing produced. This cycle is repeated until all corrections have been made and proven.

The programmer's deck (called the Source Deck) is then loaded onto the computer with a compiler program. An Object Deck (a combination of the Source Deck and the compiler program) is produced, the compiler program having inserted a number of predetermined, detailed machine instructions for any given single programmer-supplied instruction. This combined output, containing all machine instructions, is then reviewed and corrected.

The test and "debug" operations then begin, the programmer using the computer to process the Object Deck against sample data he has prepared. If there are errors, corrections are made and the program is retested. Once all corrections have been made, the system data is converted to the new system and the program is processed against "live" data. After proven operational, the program and its supporting documentation are turned over to operating personnel.

The Conceptual Program Design

An example of one way of communicating a program concept is illustrated in Figure 92. In this program "overview," the major processing steps are shown. Also, the main categories of input, output, interfaces, and files are specified. All of these, of course, would be backed up with the work sheets and mockups that specify their content, format, and volume of use.

A narrative description of the program concept should accompany the general flowchart. That description might read something like the following:

> ——.The output of this initial audit would include field error messages. However, at this point erroneous data should not be rejected. The error message would be examined by the responsible persons and correction inputs prepared. These corrections should be input to the system in the same manner as new data.——.

Designating Computer Runs

In addition to flowcharting and describing the general program concept, the systems analyst can also specify how the system is to be segmented into separate computer runs, each run constituting a single, continuous flow of a program. The conversion and validation of input might constitute one run, for instance. Others might be the sorting of data, the updating of the master file, and the processing of output.

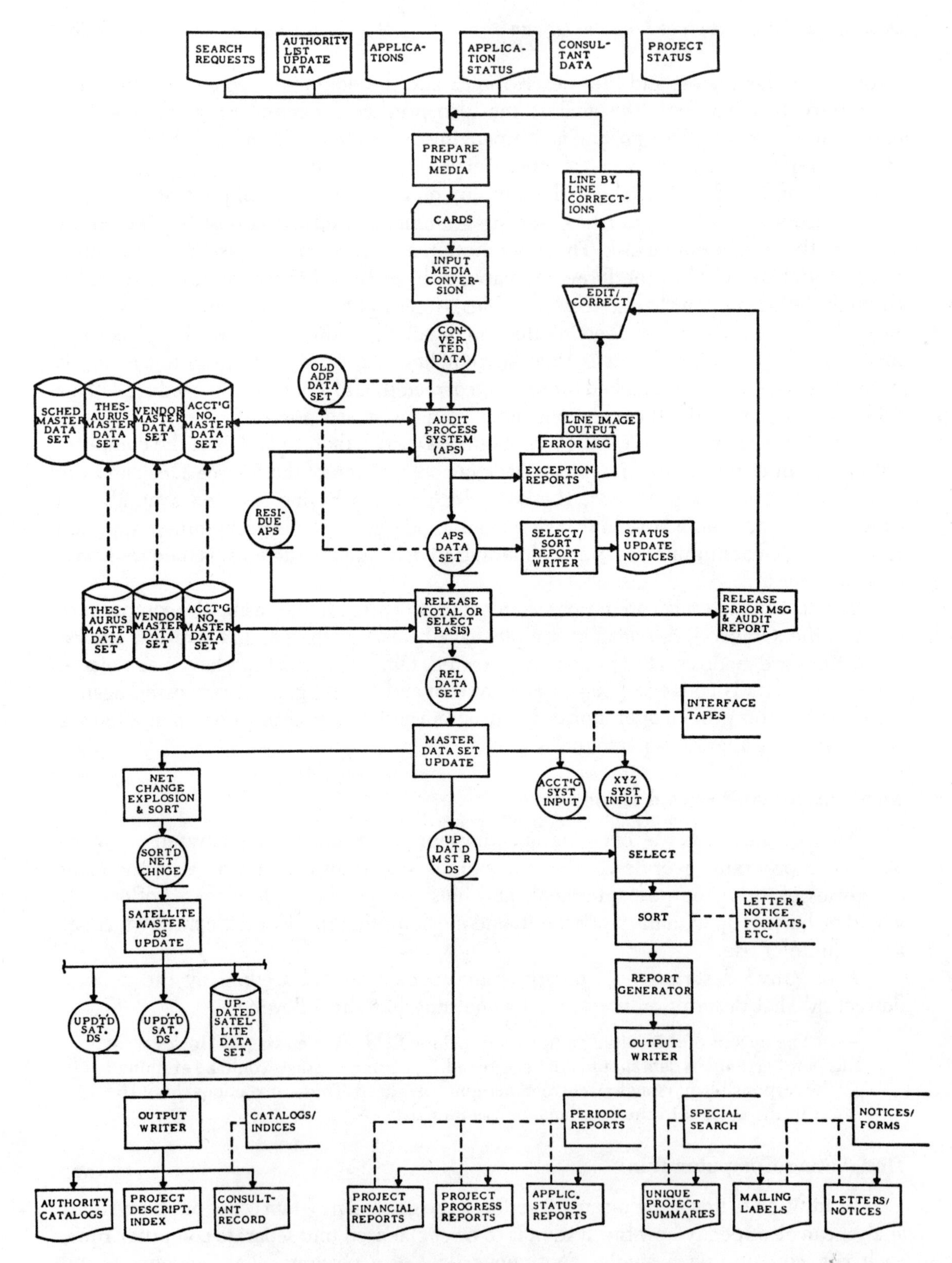

FIGURE 92: Design Example of a Conceptual Program

In segmenting the system, the systems analyst must calculate run times and costs. To do this, he must match designated equipment and alternatives to the processes involved so that comparisons can be made and optimum performance achieved.

The description of each selected run should include the calculated run time factors, the storage requirements, and a program flowchart. Each of these flowcharts should be to a finer level of detail than that described for the more general program concept flowchart.

Decision Tables

If the uniqueness of the proposed system is such that even further descriptive material is required, this detail can be provided in the form of more detailed program flowcharts, or through the use of decision tables, or both. For some situations systems analysts and programmers prefer decision tables to flowcharts.

Decision tables offer the opportunity of laying out an easy-to-visualize resolution of alternatives when there are two or more variables involved in the process. Extremely complex situations can be shown in this manner. The tables can be used independently of, or supplementary to, program flowcharts.

The basic structure of a decision table is shown in Figure 93. These tables have four sectors. The top left of the table is where the condition labels are entered, and the top right is where the condition entries are noted. The bottom of the table is where actions are expressed, their labels on the left, the action entries on the right. The table can be "read" in a manner that states, "if this condition exists (top half of the table), then this action must be executed" (bottom half of the chart).

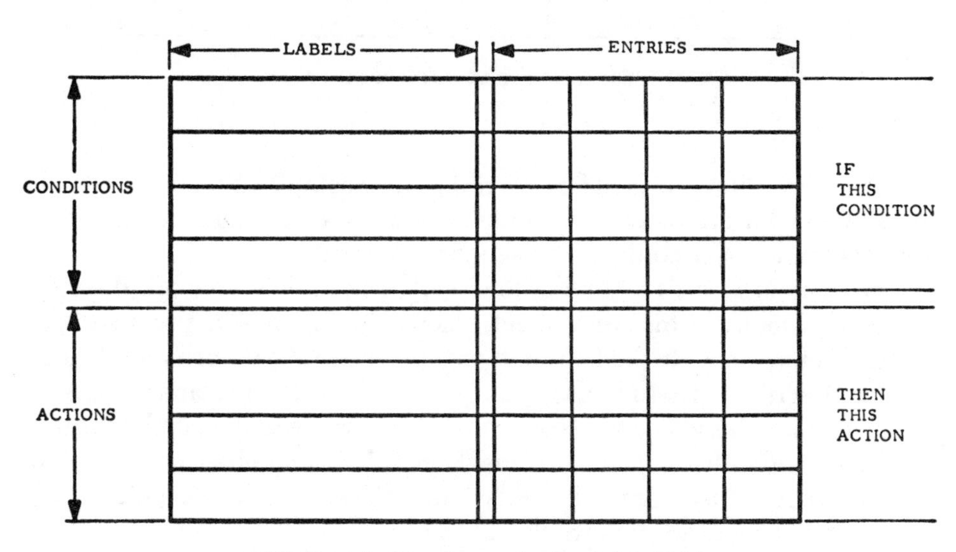

FIGURE 93: Structure of a Decision Table

There are certain conventions involved in the use of decision tables. One, for instance, is that the four sectors of the table are always separated by double lines. Another involves the uniform use of certain symbols, as follows:

Y = Yes
N = No
X = Execute
Blank = Don't Execute

Other symbols can be used but should be clearly defined in a footnote to the table. Condition labels and entries, and action labels and entries can be in the form of symbols, words, or numbers, or any combination of these forms.

An example of the use of a decision table is shown in Figure 94. Following the "rules" of the table, rule 1 shows that if the customer requests a cigar, and if cigars are in stock, and if the customer is 18 years of age or older (all of these being "conditions"), then the sale can be made (the action). Rule 2 shows that if all the conditions remain the same except the customer is not 18, then the sale cannot be executed. The rest of the table shows various other combinations of conditions and actions relative to requests for cigars and gum.

RULE	1	2	3	4	5	6	7	8
CUSTOMER REQUEST:	CIGAR	CIGAR	CIGAR	CIGAR	GUM	GUM	GUM	GUM
MERCHANDISE IN STOCK	Y	Y	N	N	Y	Y	N	N
CUSTOMER 18 OR OLDER	Y	N	Y	N	Y	N	Y	N
MAKE SALE	X				X	X		
TAKE ORDER			X				X	X

FIGURE 94: Example of a Decision Table

When decision tables are used for describing very complex situations, sets of tables are often used. One of the actions on one table, for instance, might reference another table with another set of conditions and actions.

The New-System Design section of the Workbook provides a work sheet for constructing decision tables. This work sheet, shown in Figure 95, provides space for identifying and referencing (to the flowchart or the System Flow work sheet) the system element being described. There is also space provided for noting comments. As for the spacing of the sectors, this would vary considerably according to the number of condition and action labels and entries. The work sheet provided, therefore, can serve only for some applications. As with other work sheets in the Workbook, the analyst must develop a specific form if his requirements so dictate.

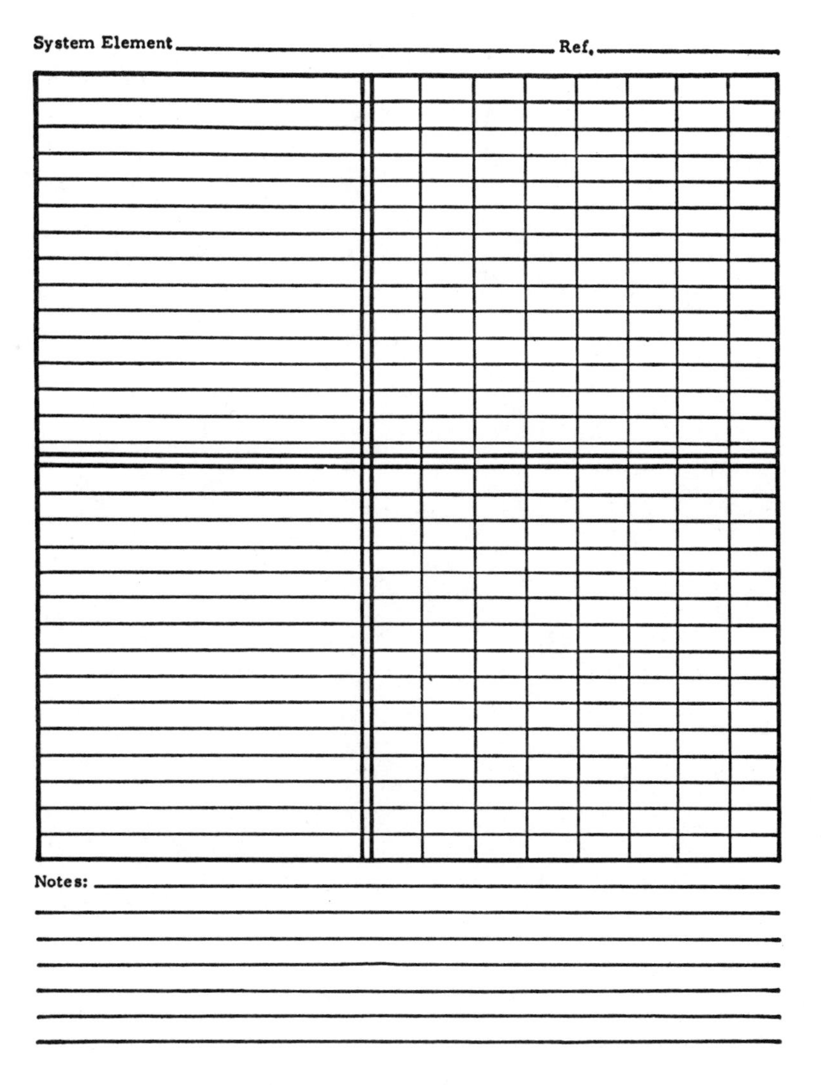

DECISION TABLE 7.10

System Element __________ Ref. __________

Notes: __________

FIGURE 95: Decision Table Work Sheet

13

Analyzing and Designing System Forms

The raw material for all systems is data; the product of its processes and its output is information. To be useful, this information must be communicated to and among the human operators in the system. Forms are the principal media for this communication because they provide a systematic and consistent method of recording and transmitting the information.

The design of the forms used in a system can contribute greatly to the efficiency and accuracy of the communication of information throughout the system. This is particularly true of data entry forms. Many of the errors in keying data can be traced to improperly designed forms. For example, if the fields on the form are not in the same sequence as the program calls for, the keyboard operator must search back and forth or up and down to find the next field. Not only does this slow the operator down, but it is also an open invitation to error.

Although not always as obvious, good forms design is equally important to efficient, accurate manual processing. The systems analyst owes it to himself to enhance the success of his system by insuring that the forms complement its operation, not hinder it.

Although many of the savings discussed will seem trivial, the individual savings have to be multiplied by the number of forms used. For instance, one minute cut from the time it takes to fill out a form, times 30 forms a day, amounts to more than 120 hours a year. Do that for 20 different forms and the savings exceed that of a full-time employee.

This kind of saving is compounded by the savings possible in handling the form in other phases of its use—filing, retrieving, transcribing data from it, making checks and audits, even in routing and distribution. Furthermore, every inefficiency built into the handling of a form invites error. If a form is difficult to identify in the file, the wrong form may be pulled. A form relating to employee Allen may be pulled instead of one relating to employee Adams. At best, time is wasted when the error is detected; at worst, wrong data are introduced into the system operation.

THE FORMS ANALYSIS WORK SHEETS

Good forms are developed through application of sound planning and analysis. The first step must be a study of the use of the form through all the phases of its life history. The most obvious process of interest is filling out the form. In addition to the handling of the form at the desk of the person filling it out, it is also normally mailed or otherwise distributed to other system operators who must extract data from it for various purposes. Copies must be filed and retrieved from files. The latter step may include storage for extended periods of time to meet legal requirements and may involve copying the filled-out form photographically.

All of these processes affect not only the layout of the form but also its construction, how it is printed and how the parts are put together.

In filling out the Forms Analysis work sheets, Figures 96 and 97, the analyst will have answered the major questions which should be asked before a form can be designed. In the process of obtaining the answers, other special circumstances may be uncovered which will also affect the ultimate design of the form or perhaps cause minor changes in system operations.

A tentative form title may be assigned. A proper title should be concise and descriptive, indicating both the function of the form and the subject of the function. Is the form to be used as an order, log, authorization, or for other purposes? These words indicate function but must be modified by a subject to clearly describe the purpose—purchase order, generating station log, trip authorization and so forth. The title must do two things to assist in the efficiency and accuracy of the system. It must insure that the person filling out the form knows, in general, what kind of information is desired and it must insure that a person wanting to do a particular task will be able to identify the proper form required.

Although the analyst should start with the assumption that a separate form is needed for each document identified on the flowchart, this may not be the case. Several documents may be combined on a single form. The analyst should first list any obvious candidates. Then, once the data items that are needed for a form are determined, he or she should go back over the Document Identification Work Sheets and look for additional candidates.

The next step is to estimate the total usage of the form, either in monthly or annual terms. This figure will affect a number of decisions as to the way a form is designed and produced. For example, some production methods, such as printing carbon snapouts or continuous forms, are feasible and economical only for large print runs. For low usage forms, a comparsion of the advantages of these techniques with the disadvantages of ordering a two or three years' supply should be made.

The potential for revision will also affect production methods and sometimes layout.

FORMS ANALYSIS **7.15**

Entering Data, Handling

Primary Ref. No. ________

Form Title ________________________________ Flow Chart Ref. __________

Is this combined form or candidate for combination?_______(List documents combined or candidates for combination on back.)

What is potential for early revision? _______High or_______Low

Usage: Total _____ per year/month and _______ per user per day

ENTERING DATA

Typewritten _____ In Office _____ Entries Made By: Employee __________

By Hand _______ In Field _______ Customer _____ Vendor __________ (Title)

By Computer: All____ or Part____; What Printer__________________

Is entire form completed by one person?_____ Or progressively, as routed _____

Source of Data (If other documents, list reference numbers): ______________

HANDLING

How are copies filed?

Limiting dimensions of file ____________________

Orientation of form ____________________

Type of file or display ____________________

If bound, type of binding ____________________

Environmental considerations ____________________

Are completed copies batched? How? ____________________

Must each copy be accounted for? ____________________

Will form be part of audit trail? _______ How identified? ______________

What special handling equipment is used:

Burster_____ Stuffer_____ Other ____________________

Folder _____ Labeller_____ ____________________

Are any of the following required or desired on form?

Procedures _______ Definitions _______ Distribution Instructions _________

Codes _______Other Special Text _______ (Attach list or draft if available)

FIGURE 96: Forms Analysis Work Sheet (Sheet 1 of 2)

There is no point in producing a lot of expensive forms if it is known that the form will be in a test status for the first six months of its use.

For usage, "per user, per day," an estimate of how many forms will be completed at a single sitting by a single user should be entered on the work sheet. This can also affect the construction of the form. If the forms are typewritten and a typist completes 25 or more at a single sitting, a continuous form similar to those used in line printers may be called for. If multiple copies are needed, the typist should not have to assemble copies and carbons each time; carbon sets or "NCR" (no carbon required) paper should be used. By eliminating the need to insert and line up a new sheet of paper or assemble a carbon

7.16

FORMS ANALYSIS

Retrieving, Extracting Data, Routing

Primary Ref. No.________

Form Title ______________________________ Flow Chart Ref. ________

RETRIEVING/EXTRACTING DATA

How will needed form be identified in file (Search Key)? ____________________

______________________________ (List data element or elements)

Is file indexed? _____ Or classified?_____ By what data element?________

Size of file: Number of documents or file-inches _____ Drawers or units_____

How often is file searched? ______________________________

If form is used for data entry:

Is it keyboarded? _____ Scanned optically? _____

Governing Program ______________________________

Prescribed data sequence? ______________________________

_________ (Indicate sequence on Data Element Design Work Sheet)

DISTRIBUTION/ROUTING

Who Needs Data?	Need Copy?	How Long Retained?	Ultimate Disposition	Makes Copies? (Process)	Received How (Mail, etc)

If copies are mailed, what type and size envelope? ____________________

What is legally required retention period? ____________________

Who will retain record copy? ______________________________

Retention media: Hard copy ____________ or Microform____________

Regulation or agency governing retention ____________________

FIGURE 97: Forms Analysis Work Sheet (Sheet 2 of 2)

set each time the typist starts a new form, the time required for the 25 or more forms might be reduced by as much as 25 percent.

On the other hand, if the forms are filled out one at a time by drivers or other field personnel, it might be beneficial to bind 25 or 50 forms in a book with a hard back and wrap-around cover for protection. In this case, carbon sets or NCR paper would not be called for without special desig If they were used, the field person would either have to remove each set from the book and find a hard surface to write on or insert a piece of thick cardboard between the top copy and the rest of the book to prevent the impression from appearing on several sets.

Entering Data

Most of the obvious inefficiencies in the use of forms are related to entering data. These also relate to the "petty" annoyances that everyone has felt in filling out forms. Fields that do not have enough space for the data called for are all too common.

The amount of space to be allowed depends on whether the form is designed for typewriter or manual completion. Most typewriters print ten characters to the inch, but for handwritten entries twice that amount of space should be allowed (or five characters to the inch). Likewise, most typewriters are fixed at six lines to the inch vertically (three for double spacing). The time it takes to fill out some forms may be doubled if the typist has to soft-roll the platen to align every line. On the other hand, a form filled out by hand in an office environment can be laid out for four lines to the inch. In a field environment where it is difficult to write, a half inch between lines should be allowed. Data squeezed into a lesser space may be illegible and detract from the accuracy of the system.

An increasing number of forms are partially completed automatically on a line printer or word processing device. In this case, the precise location of fields is critical and spacing must be compatible with the printer used.

It is also important to know whether the form is filled out by a customer, supplier or contractor or by an employee; and if by employees, what their job is. The way fields are labeled or captioned and the definitions and instructions included on the form will be determined in part by who uses the form, whether it is a secretary or manager, equipment operator or vice-president, bookkeeper or customer.

Not all forms are completed by just one person. They may be started at one station, then routed to another where additional data are entered. This is particularly true of combined documents. On purchase orders, for example, one person may enter the description, quantity and catalog number; another person, the unit price and extension; and still another, an identification of the supplier to whom the order will be sent. A copy of the order may double as a receipt of the shipment, requiring a field for the receipting signature and later as a payment authorization. The progressive nature of a form may dictate a different arrangement of data fields than would be efficient if one person filled out the entire form. Also, there may be data on internal copies that should not appear on customer or vendor copies.

The final item under this heading is Source of Data, which often dictates field sizes and may or may not influence the grouping or sequencing of fields on the new form. Form or document reference numbers should be listed if they are sources.

Form Handling

After a form is filled out, its life is not over. It is handled repeatedly, sometimes over a period of years, before it is destroyed. The amount of time wasted in handling poorly designed forms is difficult to estimate, but can be significant.

One of the big time-wasters results from not making key data items easy to locate and read. This is particularly true when a form must be located quickly in a manual file. If forms are filed alphabetically by an individual's name, that name should appear on the form where it is easy to read without excessive handling. If forms are filed right side up, as in a notebook, this usually means the upper right corner. However, in conventional file cabinets, forms are often filed sideways which calls for a different location for the file key.

In one specific case, a highly active file consisted of several thousand five by eight cards in a round, horizontal tub file that the operator rotated to bring a desired card to hand. In designing the card, it was absolutely essential to know not only the orientation of the card in the file, but also which way the operator faced in relation to the rotating tub. Otherwise, the sequential number by which the card was located might have been on the inner circumference of the file or on the side of the card away from the operator (which could still have been the upper right-hand corner of that side of the card). As it turned out, two operators used the file, sitting facing each other. This required that the sequential number appear on the outer corner of both sides (upper right on one side, upper left on the other). Other types of rotary files and rack displays create their own requirements.

On the work sheet, therefore, it is important to describe any unconventional files used. There should be a specific indication of the type of file and limiting dimensions, and a determination of the orientation of the forms in the file.

If the completed forms are bound into books, there should be an indication whether these are three-ring, prong or post binders or some other method of binding which will create unique requirements for both layout and margins. If the method used is not standard, there should be an indication of the amount of binding margins required and on which edge it will be.

Under "environmental conditions," the analyst should make note of any special condition, such as low light that might affect legibility (for example, light blue ink, often used on scanable forms, is difficult to read in either low light or bright sun). There should also be an indication of any conditions that might affect the use or permanence of the record, such as moisture or exposure to heat or vapors. (An extreme case, in this regard, involved the need for a form that could be filled out underwater by a free diver. It was for use in a program researching and monitoring the effects of power plant effluent on sea life. The various requirements of this form were met by printing a limited number of them on a special, very expensive stock which was impervious to salt water and on which the diver could write with an ordinary pencil. After they were brought to the surface, the "originals" were copied electrostatically, then erased and re-used.)

When forms are batched for transmittal or data entry, the data indicating which batch a form is to be sorted into must also be prominently displayed.

The next two items on the work sheets normally relate to the need for serially numbering each form. Numbering is usually required to keep account of each form printed or to facilitate tracing or auditing. For example, personal checks are numbered both for audit and accountability purposes. The usage, length of time between completion of a form and later reference, and the frequency of traces will determine the number of digits required to prevent duplication of numbers.

Any special equipment, such as bursters and folders, will dictate requirements for type of perforations, scoring and folding, size and sometimes for special layout elements. The precision with which labels can be attached automatically, for example, will dictate how much blank space must be left to accommodate them.

Finally, there should be an indication of any special procedures, definitions, codes or distribution instructions which should be printed on the form. Text or notes as to the contents of this material should be attached to the work sheet.

Retrieving and Extracting Data

The importance of the location of the data element by which a form is filed has been discussed earlier in this chapter. On this section of the work sheet (at the top of the second sheet), the data element which will be used for sequencing should be indicated. Usually this will be the same data element that is used to identify information needed from the file. However, searches on either of two or more data elements may be common in some files. If so, the various search keys should be indicated in the next item. This may require an index to the file or, more commonly, the file may be classified or subdivided (by date or office location, for example). If so, this should be described. The positioning and display of all these data elements should be carefully considered when laying out the form to insure their rapid location and legibility.

Sometimes there must be a trade-off between efficiency in filling out a form, with efficiency in locating it in a file, or it must be decided which of several data elements to display most prominently. Both the number of documents in the file and the frequency with which it is referenced will be factors in these decisions.

Data Entry Forms. The requirements that must be met by forms from which a keyboard operator enters data into an automated system are often more stringent than for manually processed forms. This is in part because the processing of these forms is more mechanical and less subject to in-process analysis—the keyboard operator is not usually otherwise involved with the operation of the system and is not familiar enough with the data to detect errors. It is also due in part to the fact that inefficiencies in this operation can be quantified. The keyboarder's productivity is measured in keystrokes per hour and accuracy in errors per 1,000 keystrokes.

It is important that the data to be keyed are clearly distinguished from data that are not to be keyed. It is also important that the data appear on the form in as close to the same sequence as it will be keyed. If there is a program that dictates this sequence, note of that fact should be made on the work sheet. It is also important to know what fields are of fixed length and to indicate on the form the exact number of characters allowed for these data elements.

Optically Scanned Forms. Optical scanning may also be used for data entry. Developing forms for scanning is a highly specialized field. Requirements vary with the type of equipment, the manufacturer and the model, and with the state of the art. Once the decision to use scanning has been made and the equipment specified, the analyst should insist on precise specifications for the forms to be used and should determine that the forms printer can meet these specifications.

Distribution and/or Routing

On this section of the work sheet are questions that need to be answered to determine the number of copies or parts which must be made of each form. First, by referring to the flowchart and to the previously described New-System Document Identification work sheet, a determination must be made as to who needs the data contained on the filled-out form. Do they need a copy of the form or can a copy be routed to them after it has been seen by someone else? (Or routed to someone else after they have seen it?)

"Ultimate Disposition" usually means transmittal to another organization after a short period or destruction after a longer period. If there is a legal retention requirement, there should be an indication of who is the holder of the official record copy.

Whether additional copies are made or not may dictate the use of black carbons where blue would normally be used (blue is fairly standard for handwritten forms). It may also govern the color of the copy provided to that organization. This is particularly true when a copy is microfilmed to provide an official record. For these copies red or pink paper (which will photograph black) and blue carbon, which tends to drop out of the photographic copy altogether, should be avoided.

When a copy of a form must be mailed, either by internal mail or U.S. mail, it should carry the address in a prominent place (for self-mailers) or fit into a standard envelope with or without a window. Any of these alternatives impose requirements on design and construction.

DATA ELEMENT DESIGN WORK SHEET

The next step in developing a form is to determine what data elements are required and to specify how these fields will be displayed. If the form is to serve for a single document, the list of data elements on the Data Element Design work sheet, Figure 98, can be extracted directly from the New-System Document Identification work sheet. If the form combines several documents, lists from several work sheets must be integrated.

In listing the data elements on the design work sheet, the exact nomenclature that is to appear on the form should be used. These labels or captions should be precise and unambiguous. Some common offenders of this principle are captions which read "date," "approved," "order number," or sometimes just "number." Date, for instance, can mean date received, date sent, date filled out or "today's" date. Who "approves" and what? What order number? Purchase order, work order, or job order? If these distinctions are important to the accuracy of the system, the form should be worded to insure the right data are entered in the appropriate fields.

After all the elements have been listed, key elements must be indicated—file key, index or classification term, batching key or other. Elements that must be keyboarded, and to what form (or documents) the various elements are posted, can be identified in the next two columns. This information may determine the sequence and grouping of elements in the final design.

Space is next provided for indicating any special display requirements. Key elements are often outlined with a heavy box, for example. Screening is often used to distinguish between elements to be keyed and those not keyed. In tabular matter, columns and totals which are not to be keyed or posted could be screened. (A heavier screen or pattern could also be used to block out totals or other data fields in tabular work that need not be filled in.) The technique used should be coordinated with the operator or data entry group that will be doing the keyboarding.

DATA ELEMENT DESIGN WORK SHEET **7.17**

Primary Ref. No.______

Form Title________________________________ Flow Chart Ref. ______

Data Element Name	Key Element? (for Filing, etc.)	Keyboarded?	Posted to	Type Display Desired	Definition Required?	Frequency of Use	How Entered?	Number of Characters (Max.)	Fixed Decimal?	Required on Which Parts

FIGURE 98: Data Element Design Work Sheet

In the next column on the left, an indication should be made of which elements will require a definition on the form. The necessity of a definition may depend on who normally fills out the form and his or her familiarity with the system.

Not all fields on a form are used each time a form is filled out. In the next column, the analyst should indicate the relative frequency with which a field is used (this may be a subjective term such as high, medium, low or an estimated percentage) and which are required with every use of the form. This information will be of special importance in laying out typewritten forms. Required data or fields of high frequency should be at or close to the left margin if possible. The typist will then use less time and effort in spacing across blank fields. If this placement is not possible, the high-use fields can be placed at a standard tab stop to accomplish the same end.

The analyst should next indicate how the elements are entered (typewriter, by hand, line printer, automatic typewriter, etc.) and the number of characters in fixed fields. If a fixed decimal is required, its position should be indicated.

The last column refers to multipart forms. On some forms information on one copy will not be needed or desired on another. This is often true of progressive forms where a copy is removed at one station and the remaining parts passed to another station. Purchase orders and invoices often contain information on copies retained for the originator's files that should not be transmitted to the other party. For this type of form, the analyst should indicate in the last column the part or parts on which each element will appear. (The top copy is normally part one and the remaining parts numbered from top to bottom.)

MAKING THE SKETCH

In making a sketch of a form, it is best to use a standard forms layout paper with nonprinting blue or green grids. These grids are ten to the inch horizontally and either six or eight to the inch vertically (for either typewriter or hand-entry forms).

Size

The most common size for simple forms is 8½ by 11 inches. Standard sheets of paper of the type used for printing most forms are 17 by 22 inches from which four 8½ by 11 inch sheets can be cut without waste. Therefore, even portions of either 17 or 22 inches should be used for dimension purposes, whenever possible. (Note: card files are normally 3 x 5, 4 x 6 or 5 x 8 inches and many government files are 8 x 10½ inches, none of which can be cut from standard stock without waste.)

Continuous forms, carbon snapouts and some other types of multipart forms have a perforated stub along one edge of the form, to hold the sets together, which is usually ¾ of an inch wide. However, standards have evolved that allow the "torn-out" size of these forms to be the same as for forms printed on sheets.

Margins

The analyst should next establish the margins required. If there is a binding edge, he or she should indicate which edge it will be and allow sufficient margin as required by the binding method.

Another consideration that will determine one of the margins is the requirement for "gripper" space. This is space that must be free of type or rules to allow for the mechanical operation of the press. It varies with the type of press, but is usually either ⅜ or ¼ inch.

Style

Although the word "style" has connotations relating to aesthetics and personal preference, considerations of efficiency and system operations also affect the style of a form.

There are several basic alternatives for laying out individual data elements on a form (e.g., boxed or open). (See Figure 99, Example of Styles.) The boxed style, with captions

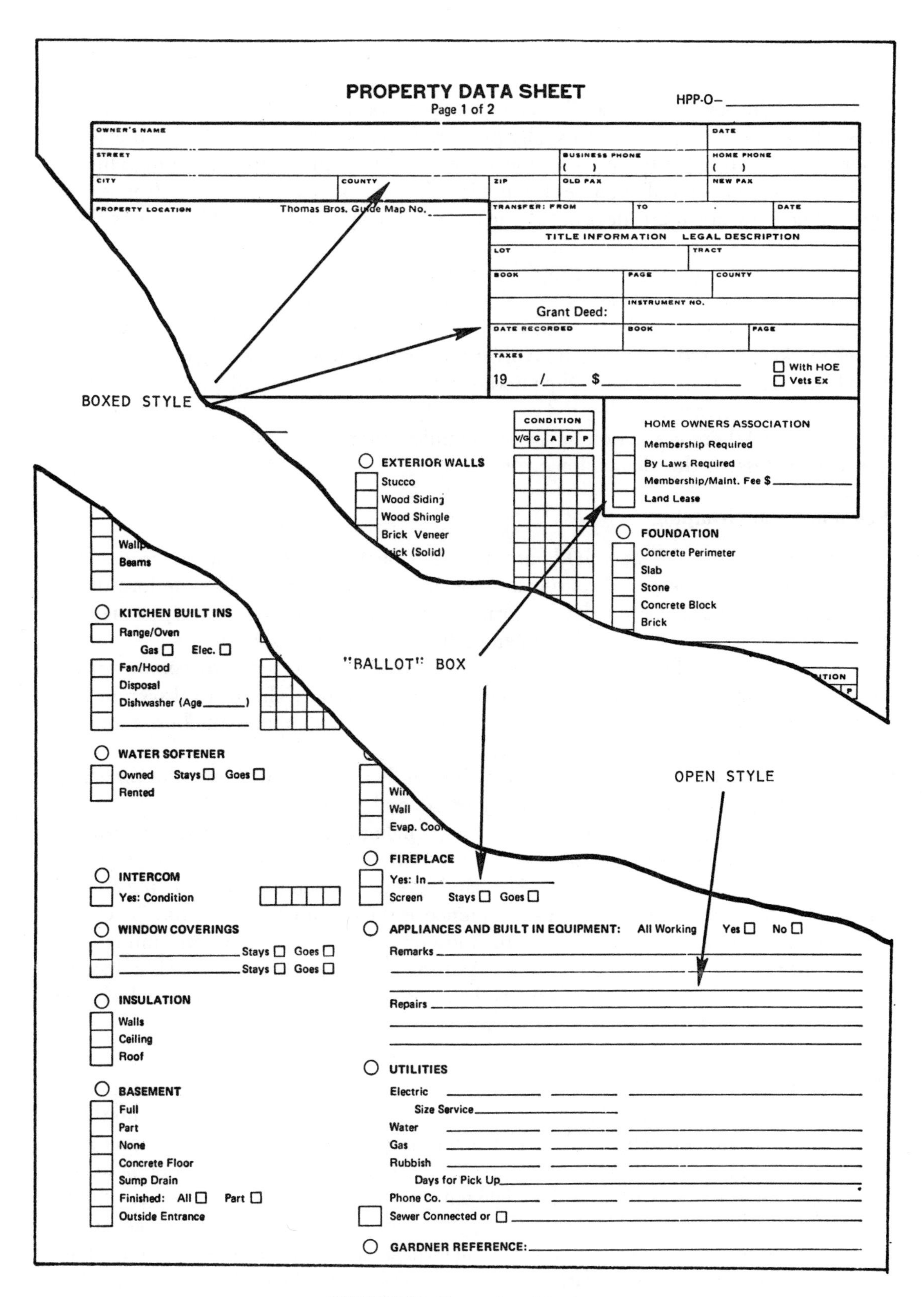

PROPERTY DATA SHEET
Page 1 of 2
HPP-O– ______

OWNER'S NAME			DATE
STREET		BUSINESS PHONE ( )	HOME PHONE ( )
CITY	COUNTY	ZIP / OLD PAX	NEW PAX

PROPERTY LOCATION Thomas Bros. Guide Map No. ______

TRANSFER: FROM | TO | DATE

TITLE INFORMATION LEGAL DESCRIPTION

LOT		TRACT
BOOK	PAGE	COUNTY
Grant Deed:	INSTRUMENT NO.	
DATE RECORDED	BOOK	PAGE

TAXES
19____/____ $______ ☐ With HOE ☐ Vets Ex

CONDITION
V/G G A F P

○ EXTERIOR WALLS
☐ Stucco
☐ Wood Siding
☐ Wood Shingle
☐ Brick Veneer
☐ ick (Solid)

HOME OWNERS ASSOCIATION
☐ Membership Required
☐ By Laws Required
☐ Membership/Maint. Fee $______
☐ Land Lease

○ FOUNDATION
☐ Concrete Perimeter
☐ Slab
☐ Stone
☐ Concrete Block
☐ Brick

☐ Wallp
☐ Beams

○ KITCHEN BUILT INS
☐ Range/Oven
Gas ☐ Elec. ☐
☐ Fan/Hood
☐ Disposal
☐ Dishwasher (Age______)

○ WATER SOFTENER
☐ Owned Stays ☐ Goes ☐
☐ Rented

☐ Wi
☐ Wall
☐ Evap. Coo

○ FIREPLACE
☐ Yes: In______
☐ Screen Stays ☐ Goes ☐

○ INTERCOM
☐ Yes: Condition

○ WINDOW COVERINGS
☐ ______ Stays ☐ Goes ☐
☐ ______ Stays ☐ Goes ☐

○ INSULATION
☐ Walls
☐ Ceiling
☐ Roof

○ BASEMENT
☐ Full
☐ Part
☐ None
☐ Concrete Floor
☐ Sump Drain
☐ Finished: All ☐ Part ☐
☐ Outside Entrance

○ APPLIANCES AND BUILT IN EQUIPMENT: All Working Yes ☐ No ☐
Remarks ______
Repairs ______

○ UTILITIES
Electric ______
Size Service ______
Water ______
Gas ______
Rubbish ______
Days for Pick Up ______
Phone Co. ______
☐ Sewer Connected or ☐ ______

○ GARDNER REFERENCE: ______

FIGURE 99: Example of Styles

at the top of the box, emphasizes the data, clearly delineates each field and also allows the most efficient use of space. It is usually preferred unless other requirements dictate an open style. It avoids forcing a typist to space through the captions to start typing a line. It also allows easier vertical alignment of fields. All of this saves valuable keystrokes. The open style provides more flexibility, particularly for open-ended fields such as those labeled "Remarks" or "Comments."

Another approach involves the use of a tabular style. This is an efficient method of recording some types of data and allows maximum use of space. "Ballot Boxes" are useful when the alternatives are fixed, precise and can be indicated efficiently with a word or check mark. Still another approach involves the need for accommodating questions with long blanks for answers, blanks in the middle of the text and other similar requirements.

It is not necessary to be limited to a single style on one form. Any and all styles may be combined if there is a reason to do so. But unless there is a valid reason for variation, consistency lessens the chance for error.

Sequence and Grouping

In determining the way fields are arranged on the form, it should be decided whether grouping is required, either for efficiency or in order to make the form more understandable. For instance, data on forms that are filled out progressively as they are passed from station to station should generally be grouped by station.

Also, when some but not all of the data on a form is to be entered into a computer data base, the information to be keyed should be grouped together. There are also groupings dictated by convention or logic, such as name, address and telephone number, or debits, credits and balance.

On data entry forms the sequence may be rigidly dictated by a computer program. On other types of forms there will usually be several options to consider.

One set of options that is important in sequencing data fields is related to posting. Greatest efficiency in *filling out* the form calls for putting data fields in the same sequence as on the source document. On the other hand, greatest efficiency in *extracting data* to post to another form requires the same sequence as the document posted to. The choice will depend on which operation is done the most often, is the most time consuming, or is done under the most pressure. Accuracy in posting is also a factor. Accuracy is enhanced when the sequence on both documents is the same and the choice may go to whichever posting sequence (to or from this form) is most critical to system accuracy.

Screening

Sometimes it is desirable to subdue some areas of the form or printed elements (lines or text). This is often done by screening, which is used either to lay a light background of very smalll dots over an area or to break up otherwise solid elements into a dot pattern.

Screening is particularly useful on data entry forms where it may be used to subdue any fields not to be keyed. Or, attention may be called to the fields to be keyed by screening the labels and leaving the rest of the form clear. Many data entry organizations have established conventions relative to screening and it could be useful for the analyst to check with the operators who will be working with the forms as to their rules and preferences.

Line Weights

After the analyst has completed his pencil draft, he should indicate any particular requirements for heavy or light lines. He won't be expected to specify line weights by point size since that is the job of the designer or artist, but he should indicate his preferences.

There are two types of light lines to be aware of. Screened lines are the lightest and are often used to separate character spaces in fixed fields on data entry forms. They can be used where a very light guideline is needed or total separation of spaces is not desired.

Hairlines provide a slightly stronger separation between spaces (for example, to separate pairs of characters for year, month and day within fixed, six character date fields).

Instructions

Generally, few instructions are required on the form itself. When they are required, they should be as concise as possible and in outline or step-by-step format rather than narrative. They should also be placed above the part of the form to which they apply, whenever possible. This is to increase the probability they will be read before the form is filled out improperly.

Other Elements

If the form is to be perforated, prepunched for binding, scored to facilitate folding, or serially numbered, the analyst should indicate the location of these features on the sketch. Also, if any address or other information must show through a window envelope, the precise location and dimensions of the window should be indicated.

FORM CONSTRUCTION

The simplest type of forms construction is the single sheet form. However, there are some basic options available for more complex forms which can affect the efficiency and accuracy of the system. Since the capabilities of printers for producing these varies, the analyst should make sure as to what specific options are available.

Multipart Forms

This refers to forms of several sheets or parts on which the data are entered on the top sheet and impressed on subsequent sheets through the use of either carbon or NCR (no carbon required) paper. The same thing can be accomplished by assembling single sheet forms and carbon paper each time the form is filled out (or making electrostatic copies), but this should be used only when the volume is low and the cost of multipart forms is not justified.

Carbon Snapouts

A multipart form in which the carbon is bound with the parts is called a snapout if all the carbons can be removed with a single operation. Snapouts and other construction methods in which the carbons are integrated should be used when the individual forms are separated from the pad or book in which they are bound before they are filled out. An alternate to snapout construction involves the use of a carbon layer on the back of each part. Applications of this technique are the same as for NCR paper.

Spot and Pattern Carbon

Sometimes only some of the data is needed on parts other than the top sheet. In this case, spots of carbon can be applied to the back of the sheet. Patterns may also be used for larger areas.

Continuous Forms

These forms are printed on a continuous sheet of paper. The individual forms are separated from each other by perforations and may have multiple parts with integral carbon. Whenever data are entered by a line printer, continuous forms must be used.

14
Determining the Equipment Configuration

As a parallel activity to designing the new system the analyst must determine, then specify, the system's required equipment configuration. The selection must be made on the basis of technical and economic feasibility. The broadest level of consideration involves:

- Using Existing Equipment
- Leasing or Buying New Equipment
- Using a Data Processing Service Center
- Using a Time-Sharing Service
- Using a Distributed Processing Concept

The direction and approach the analyst must take in his evaluation will vary significantly according to these broad considerations. It may be, for instance, that the nature of the system being studied limits the analyst to the use of the existing equipment. This would be the case if the study area was limited in scope, the system being quite small in relationship to other operating systems utilizing the same equipment. If this was the case the analyst would have to tailor his design to fit the equipment, evaluating the equipment on the basis of its capacity, in view of other workloads, to handle the storage, processing, and available-time requirements of the new system. If equipment capacity was small, and if management was unable or unwilling to replace or augment it, the analyst might well find himself in the position of cutting the foot to fit the shoe—trimming the system's capability to fit the existing equipment's capacity.

If the nature of the system is such that the other approaches should be considered (leasing or buying new equipment, or utilizing a service), then the analyst must evaluate the alternatives, documenting their features in a manner so they can be readily compared in relationship to the proposed system's requirements.

Equipment requirements derive directly from system functional requirements. It is the purpose of this chapter to describe a systematic way of utilizing work sheets to translate these requirements into a specific hardware configuration.

In considering equipment, there are several categories of equipment that might be essential to the operation of a system. For automated systems, of course, the selection of computing and perhaps teleprocessing equipment is critical. However, there may be important equipment decisions in the design of wholly manual systems as well.

In addition to computer and teleprocessing equipment, there is communications equipment of all types, reproduction equipment, office machines, and other such devices which may apply to either type of system. The work sheets described in this chapter, and provided in the New-System Design section of the Workbook can, and should, be used to document and compare all such hardware requirements.

Another factor that must be considered is that all design is an iterative process and involves compromise between requirements and solutions. This is particularly true in the task of matching design criteria to equipment capability. One operation may indicate a requirement for speed, another for capacity. Considering the dollars available or practical for equipment acquisition, both factors may not be available in the same machine. One or the other, or both, of the requirements must then be modified. The designer will find it necessary to repeatedly re-examine both the design concept and the equipment prospects, searching for some alternative that will soften the compromises required.

In gathering information regarding equipment capabilities so that comparisons can be made, the analyst will find that such information is not hard to come by. The problem is often the opposite. There is so much material available, albeit mostly promotional and disorganized, that it is often difficult to sort out the useful from the valueless. With these conditions, a useful alternate can be the comparison charts published periodically by many of the trade journals. Also available are copyrighted guides available on a subscription basis.

Two types of cost estimates are derived from the work involved in determining the proposed system's equipment configuration. One is the equipment operating costs which serve as input to the task of estimating all operating costs (covered in the next chapter). The other is the equipment acquisition and activation costs. This latter information serves as input to the implementation planning task where there costs are summed with all other implementation costs (covered in Chapter 17).

HARDWARE/PROGRAMMING OPTIONS

A preliminary step in determining a system's equipment configuration involves a review of the hardware and software options that are available to the analyst.

Types of Input

There are a variety of input options, including those, of course, that are suitable for manual data entry. Of those used in mechanized systems, the most familiar is the punched card. In addition to this, there are input devices for reading mark-sense cards, optical character recognition devices, magnetic ink character recognition devices and keyboard terminals.

Data Storage Media

The items of equipment used for the filing of material in manual systems are familiar to all: vertical file cabinets and shelves, visible files, open files, tub files, rotary files and wheel files. There are also special cabinets used for storing the materials used in mechanized systems. These include cabinets for storing magnetic tapes, punched cards and mark-sense cards.

Film is an important storage medium in many systems. Methods include rolled microfilm, microfilm magazines, microfiche and film inserted into sortable punched cards (aperture cards).

There are four main categories of computer storage media: magnetic tape storage, magnetic disk storage, magnetic drum storage and data cell storage.

Teleprocessing Systems

Some systems require communication links for the purpose of transmitting data from one location to another. The most common means for accomplishing this is through the use of telephone lines. (Alternate options: microwave circuits and private wire circuits.)

Types of Output

Several types of output options can be exercised depending on the needs of the system. First, hard-copy output can be generated using typewriters, on-line typewriters, line printers and various types of plotting devices. Machine-readable output can be generated as well, including punched cards, magnetic tape or paper tape. Microfilm can also be produced through the process of generating machine-readable output, then translating it into a display which is photographed. This is called Computer Output Microfilm (COM).

Several types of soft-copy output can be generated, again depending upon the needs of the system. One type is an audio response output. The other is a visual response which displays graphic or alphanumeric materials on a video screen.

Program Languages

There are options available to the system designer relative to the language to be used to communicate with the equipment and to direct operations. Several of the more commonly used languages are listed and described below.

COBOL, from *CO*mmon *B*usiness *O*riented *L*anguage. Closely resembles ordinary English. Requires large computer.

FORTRAN, from *FOR*mula *TRAN*slating System. Best suited for scientific and mathematical applications. Can be used for some business system applications. Can be run on large or small computers.

BASIC, from *B*eginners *A*ll-Purpose *S*ymbolic *I*nstruction *C*ode. Closely resembles ordinary English. Easy to learn and apply. Used in many small computer and time-sharing applications.

RPG, from *R*eport *P*rogram *G*enerator. Easy to use. Available on small and large computers. Widely used in business applications.

Assembler Language. Only language available on many small machines. Uses system of abbreviations which is highly efficient in its use of computer's primary storage area.

EQUIPMENT REQUIREMENTS AND EVALUATION WORK SHEET

The key work sheet provided for matching system needs with equipment, is the Equipment Requirements and Evaluation work sheet shown in Figure 100. In using this work sheet, the system should first be segmented, as described in the previous chapter, into logical runs. Next, each run should be further segmented into logical operational steps that can be readily identified with a single type of equipment. If more than one type of equipment is directly associated with a single operational step, then separate work sheets should be used. For cross-referencing purposes, these operational steps should be identified with the reference number originally assigned on the System Flow work sheet.

After the segment and operational step have been identified on the work sheet, the quantity or volume factor for each separate run (or transaction) should be noted in terms of both peak and average quantities, followed by an indication of frequency. Any special requirements, such as file capacity, are next noted, followed by an identification of the equipment type and quantity of units required. Since capacities of equipment vary widely depending on make and model, this first indication of quantity of units required must necessarily be only a rough estimate to be used as a guideline.

The next section of the work sheet provides space for identifying the likely source of this equipment. If it is existing equipment, then an indication is made as to its capacity and availability as related to this application. If the equipment type is not available, or if its remaining capacity is not suitable, then the analyst indicates his current thinking as to whether the requirement should best be met by leasing or buying equipment, or by subscribing or contracting for one of the several forms of service that is available.

The work sheet then provides space for recording and comparing three alternate approaches to the equipment. (If more alternatives are desired, supplemental sheets can be used.) If the analyst wishes, one of the models thus identified can be the existing equipment which can thus be compared from a performance standpoint with several alternatives. Descriptive data regarding the existing equipment can be derived from the Equipment Description work sheet located in the System Environment Factors section of the Workbook.

Each alternative's estimated performance as it relates to the subject application is then noted. This is in terms of the estimated processing times per run (or transaction), per week, month, or year. Costs are also estimated on the same basis, with the method for determining the cost rate specified (actual charge, or bid rate, or equipment acquisition costs prorated plus operating costs, etc.). Finally, significant equipment characteristics and limitations (including capacity) are noted.

EQUIPMENT REQUIREMENTS AND EVALUATION 7.11

Sheet ____ of ____

System Segment ________________

Operation Step ________________

________________ System Flow Ref. No. ________

____

Vol. Factor per Run (or Transaction): Peak ________ Aver. ________ Freq. ________

Special Requirements ________________

Equipment Type/Quantity Required ________________

Source: Existing? ________ Capacity Suitable? ________ Available? ________

Lease? ________ Buy? ________ Service? (Specify Type) ________________

Alternate # ____

Make, Model, & No. ________________ Quantity ________

Est. Processing Time: Per Aver. Run ________ Per Mo. ________

Cost Rate Per Run ________________

Method of Determining Rate ________________

Est. Costs: Per Run ________ Per Mo. ________

Significant Characteristics & Limitations ________________

____

Alternate # ____

Make, Model, & No. ________________ Quantity ________

Est. Processing Time: Per Aver. Run ________ Per Mo. ________

Cost Rate Per Run ________________

Method of Determining Rate ________________

Est. Costs: Per Run ________ Per Mo. ________

Significant Characteristics & Limitations ________________

____

Alternate # ____

Make, Model, & No. ________________ Quantity ________

Est. Processing Time: Per Aver. Run ________ Per Mo. ________

Cost Rate Per Run ________________

Method of Determining Rate ________________

Est. Costs: Per Run ________ Per Mo. ________

Significant Characteristics & Limitations ________________

____

Comments ________________

FIGURE 100: Equipment Requirements and Evaluation Work Sheet

The bottom of the work sheet provides space for general comments relative to the basic requirement and all alternatives.

EQUIPMENT UTILIZATION SUMMARY WORK SHEET

The preceding work sheet is used for specifying an equipment-type requirement and comparing alternatives on the basis of a single operational step of the system. The Equipment Utilization Summary work sheet, shown in Figure 101, relates each type of

EQUIPMENT UTILIZATION SUMMARY **7.12**

Selected Equipment	No. Units	Capacity Available For This Application	Operational Step	Equipment Capacity Required

FIGURE 101: Equipment Utilization Summary Work Sheet

equipment to all operational steps so that cross-uses (multiple uses) of equipment types can be determined.

The first three columns are used for recording each type of equipment selected as the best alternative from the Equipment Requirements and Evaluation work sheets. This can be existing equipment or proposed new equipment. The equipment is identified in the first column, and the number of units required to accommodate all system requirements is entered in the second column. The third column is used for recording the most appropriate measure of available capacity so that this measurement can be directly related to the requirements of the new system. This can be in units of information, time,

7.13

NEW-SYSTEM EQUIPMENT CONFIGURATION

Sheet____of____

Application __________

General Description of Configuration __________

Equipment Type, Make, Model, & No.__________ No. Units______

Source: Existing?______ Lease?______ Buy?______ Service?______

Notes Regarding Source__________

If Not "Service", % Usage: This Applic., Peak______ Aver.______

Dedicated for Other Use______ Balance Avail.______

Total Cost/Charge______ % Chargeable to This Applic.______

Availability/Delivery Factors__________

Special Characteristics, Advantages__________

Equipment Type, Make, Model, & No.__________ No. Units______

Source: Existing?______ Lease?______ Buy?______ Service?______

Notes Regarding Source__________

If Not "Service", % Usage: This Applic., Peak______ Aver.______

Dedicated for Other Use______ Balance Avail.______

Total Cost/Charge______ % Chargeable to This Applic.______

Availability/Delivery Factors__________

Special Characteristics, Advantages__________

Equipment Type, Make, Model, & No.__________ No. Units______

Source: Existing?______ Lease?______ Buy?______ Service?______

Notes Regarding Source__________

If Not "Service", % Usage: This Applic., Peak______ Aver.______

Dedicated for Other Use______ Balance Avail.______

Total Cost/Charge______ % Chargeable to This Applic.______

Availability/Delivery Factors__________

Special Characteristics, Advantages__________

FIGURE 102: New-System Equipment Configuration Work Sheet

or percentage. Since there can be multiple units of the same type of equipment, the capacity measurement for a single unit must be multiplied by the number of units specified in order to have a figure that represents total capacity available for the proposed application. This also implies that this figure must be further modified if a portion of the equipment's capacity is dedicated for use for some other system outside the scope of the proposed application.

The two right hand columns of the work sheet provide space for recording each separate operational step and the corresponding requirement for equipment capacity. In this manner the operation steps are grouped according to equipment type. If two or more operational steps require the use of the same type of equipment, then their separate capacity requirements can be summed and equated with the available capacity.

NEW-SYSTEM EQUIPMENT CONFIGURATION WORK SHEET

After various alternatives have been evaluated, and the best equipment utilization determined, the selected units of equipment are assembled into a proposed equipment configuration. The New-System Equipment Configuration work sheet, shown in Figure 102, can be used for documenting this selection.

The application (proposed system) is identified at the top of the work sheet, followed by a brief description of the overall equipment configuration. Each work sheet then provides space for describing three of the selected types of equipment that contribute to making up the total configuration. Each type is first identified, and its make, model and model number, and number of units required specified. The source of the equipment is indicated in terms of whether it is existing equipment, or to be leased, or purchased, or if a service for its use is to be contracted for. There is also space provided for noting comments regarding the equipment source, such as identifying equipment or service vendors.

Notations regarding the equipment's utilization and cost are recorded next on the work sheet. As to cost, this factor should be recorded in terms of total costs or charges, and what percent of that cost is chargeable to the proposed application in the event that there's a difference. A comment as to the equipment's availability and/or expected delivery schedule should be recorded next on the work sheet.

Finally, special characteristics and advantages of this equipment should be noted. If it is to be purchased or leased, manufacturer's descriptive material and specifications can be attached to the work sheet.

SITE PREPARATION AND ACTIVATION CONSIDERATIONS WORK SHEET

If all or part of the equipment required for the new system is to be purchased or leased, the Site Preparation and Activation Considerations work sheet (Figure 103) should be used. Site selection factors should be noted, and a tentative floor plan specified, with a rough draft of that plan appended to the work sheet. If there are special problems or considerations regarding the structure or floor loading, these are to be noted, as well.

The work sheet next calls for brief notations as to factors concerning air conditioning, electrical and cable requirements, safety and security, and schedule. The planned approach to converting from the old to the new equipment should be noted, as well as an idea as to the disposition of existing equipment that would be made obsolete by the new equipment.

The bottom part of the work sheet provides space for noting anticipated requirements as they apply to equipment room furnishings and supplies.

7.14

SITE PREPARATION AND ACTIVATION CONSIDERATIONS

Site Selection Factors ____________________

Tentative Floor Plan (Attach Rough Draft) ____________________

Structural & Floor Loading Considerations ____________________

Air Conditioning, Electrical, & Cable Factors ____________________

Safety & Security Factors ____________________

Schedule Factors ____________________

Cutover/Conversion Strategy ____________________

Disposition of Existing Equipment ____________________

Furnishings Considerations ____________________
- Files, Storage Cabinets, Carts ____________________
- Communications Equipt (Attach Supplemental Description) ____________________
- Desks, Chairs, Work Tables ____________________
- Other ____________________

Supplies Considerations ____________________
- Printed Forms & General Supplies ____________________
- Mag. Tapes ____________________
- Disk Packs ____________________
- Other ____________________

FIGURE 103: Site Preparation and Activation Considerations Work Sheet

DECENTRALIZED SYSTEMS

Back at what seems to many like the dawn of history, most people in any organization processed their own data. The computer came along and offered a way to do this work faster, cheaper and with a lower incidence of error. Because of the sophistication of the equipment and its high cost, it was best to do this work on a single, large, centrally located computer. In practice, the organizational unit in need of this service "batched" the data, sent it in for processing, and waited for the output. In time, a few applications were provisioned with remotely located terminals which provided direct access to the centralized computer.

As time went on the trend towards centralization slowed, stopped, then reversed. This was a result of rapid technological advancements. Computer hardware designers were finding it increasingly feasible to cram ever-greater capabilities into ever-smaller packages, and all of this at a tremendous reduction in cost. The resultant "minicomputers" hastened the era of decentralization.

Minicomputers

The system designer, as one of the options in his design effort, must consider the feasibility of using minicomputers. Systems that use this type of equipment have been labeled as distributed computing systems, distributed processing systems, distributive data processing systems, network computing systems and decentralized computer systems. No matter what the name is, the idea has been to return to the user, where possible, those functions the user should have. Furthermore, there has been a growing need to provide computing capability out in an organization's work areas.

Examples of Applications

Practical applications of decentralized processing systems are those where the equipment is exclusively dedicated to a specific user, and where the workload is readily contained within the equipment's capacity.

Typical of systems that are applied in a decentralized mode are those that fall in the categories of order processing systems, inquiry systems, word processing systems, management information systems, personnel systems and others. Small and medium-size industrial firms that, in the past, could not afford computers, now enjoy the convenience and utility of distributed processing systems for such applications as:

- Inventory Control
- Product Data Control
- Master Scheduling
- Material Requirements Planning
- Purchasing Control
- Work Order Control and Tracking
- Assembly Withdrawal and Kitting Control
- Sales Order Entry-Shipment-Billing
- Sales Forecasting
- Shop Scheduling and Loading
- Job and Labor Reporting
- Cost Accounting
- General Accounting
- Cost and Price Estimating
- Capacity Planning
- Tool Control

- Field Spares Planning
- Lot Traceability

Advantages of Decentralized Processing Systems

A minicomputer hardware system can consist of a keyboard terminal and related input devices, a visual display unit, computing and memory hardware, and output devices of different types. Systems can be designed so that when entering a specific type of data, or making a query, an operator can follow step-by-step instructions that are displayed on the video screen. This eliminates the need for manuals of written instructions. Furthermore, applications can be designed so that when data are entered, there is an automatic check for accuracy.

A minicomputer system located "on-site" where it is needed, constitutes a system that is highly responsive to the needs of the user. It is flexible, comparatively easy to program, will support a wide variety of applications, and with visual display capabilities is easy to use.

Since data is entered at its source, it is entered by the people who are most familiar with it. These are the ones who can readily recognize bad data (incorrect part numbers, for instance) and make on-the-spot corrections.

The systems that are designed for this type of equipment are interactive, permitting a dialogue between the operator and the computer. They provide a "now" type of environment where the user can find out what's happening at that very moment in time. In addition, there are multiprocessing capabilities which enable the simultaneous running of various programs using various languages.

The user of such a system, no longer dependent on a centralized system, can exercise closer control over such matters as priorities and the management of backlog. An added advantage is that management reports can be generated locally and quickly, as needed.

Problems with Decentralized Processing Systems

A recurring problem with decentralized processing systems is a tendency for them to become undedicated. Their original purpose begins to become submerged through the addition of applications and periphery equipment to accommodate other users. They soon begin to resemble complex centers, suffering the same problems that have plagued centralized data processing operations from the beginning.

There is a problem, too, with hardware standardization. A great variety of ingenious hardware is available. The problem surfaces when attempts are made to combine several existing but separate systems into a single one. If the hardware is not compatible, the merging process can be costly.

Comparing Equipment

A work sheet for comparing the features of candidate equipment for a decentralized data processing system is displayed in Figure 104, Distributed Processing System Equipment Comparison Chart. Information relative to the equipment of three different manufacturers can be compared on this particular work sheet. Several key questions on

DISTRIBUTED PROCESSING SYSTEM **7.18**

EQUIPMENT COMPARISON CHART

APPLICATION DATA

Application ________________ No. of Stations ____

Stand-Alone Installation? ________ or On-Line to Host CPU? ____

Remote Data Entry? ________ or Source Data Entry? ____

Interactive Processing? ________ or Batch Processing? ____

Other Similar Applications Within Organization? ________ Where? ____

Equipment Used ________________ Need to Standardize? ____

COMPARISON DATA	1	2	3
Manufacturer			
Model Name/No.			
Purchase Price - Basic Unit			
Peripheral Prices:			
1.			
2.			
3.			
4.			
5.			
Software Price			
Software Conditions			
Total Price/Unit			
Lease Costs/Unit			
Input System			
Storage System			
Storage Capacity			
Type of Output			
Output Speed			
Operator Training Time			
Order Lead Time			

FIGURE 104: Distributed Processing System Equipment Comparison Chart

the form deal with the problem cited in the previous paragraph. They ask for an identification of other decentralized data processing applications that may be extant within the organization, and for an indication of whether or not there is a need for standardization.

Blank spaces are left on the form for the analyst to identify the specific type of peripheral equipment that would be needed on this application. Under "Software Conditions," the analyst would indicate whether it was an off-the-shelf package, a customized package or whatever.

A comparison of small computers can be made using the work sheet shown in Figure

7.19

SMALL COMPUTERS

COMPARISON OF HARDWARE ALTERNATIVES

APPLICATION ______________________

	1	2	3
Manufacturer			
Model Name/Number			
Purchase Price-Basic Unit			
Monthly Lease-Basic Unit			
System Language			
User Language			
Types of Processing			
1.			
2.			
3.			
4.			
5.			
6.			
Software Packages Required			
1.			
2.			
3.			
4.			
5.			
6.			
7.			
8.			
Input/Output Devices			
1. Keyboard/Printer			
2. Punched Card			
3. Magnetic Card			
4. Paper Tape			
5. Magnetic Tape			
6. Mag Disc or Drum			
7. Line Printer			
8. CRT			
Storage Media			
Storage Capacity			
Storage Access Time			
Vendor Services			
1. Training			
2. System Design			
3. Programming			

FIGURE 105: Small Computers—Comparison of Hardware Alternatives

105, Small Computers—Comparison of Hardware Alternatives. Spaces are provided on this form for entering the processing needs for this specific application (such as multiprogramming, remote or direct inquiry, etc.) and for listing the software packages required (such as order entry, payroll, inventory control, etc.)

15
Defining the Proposed System's Value

The system design effort is an iterative process of designing the system, testing its values, then redesigning to minimize costs and optimize the value of its benefits. As described in the previous chapter, part of this involves the process of determining the optimum equipment configuration. This chapter deals with developing estimates of the total system's operating costs and benefits, projecting these factors, and comparing values against the existing-system's cost/benefit baseline.

To aid in this appraisal of system values, seven work sheets are provided in the Evaluation Criteria section of the *Workbook*. The first is identical, except for the title, to the Cost Breakdown By Process Step work sheet presented earlier in Chapter 9. The next two are used in defining the proposed system's operation tasks and the resources required for their accomplishment. The remaining four are used in comparing values with the existing system in terms of:

- Operating Costs and Benefits
- Life-Cycle Costs
- Impact on Profit and Cash

The last four work sheets would already contain, at this point in the system's analysis, the measurements of the existing system's values, summarized from the work sheets described in Chapter 9.

PROPOSED SYSTEM COST BREAKDOWN BY PROCESS STEP

An example was given in Chapter 9 of a simple process for responding to price queries received in the mail. A flowchart of the existing system was shown, together with a graph that clearly illustrated that the workload would soon exceed the existing system's capacity. The costs of the existing system were summarized on an Existing-System Cost Breakdown by Process Step work sheet, resulting in a unit cost factor of $2.45 per answer.

A flowchart of a possible system improvement concept for that task is shown in Figure 106, Flowchart of Proposed Price Query Answering System. In this approach, the

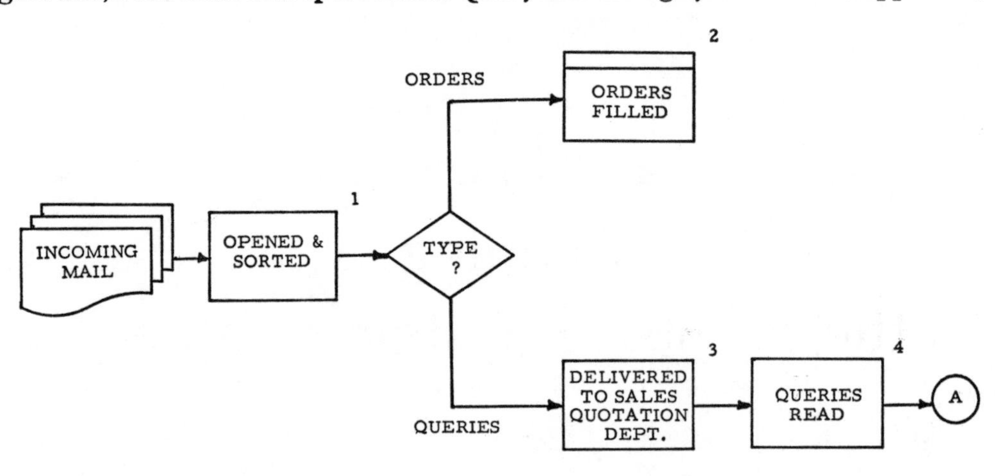

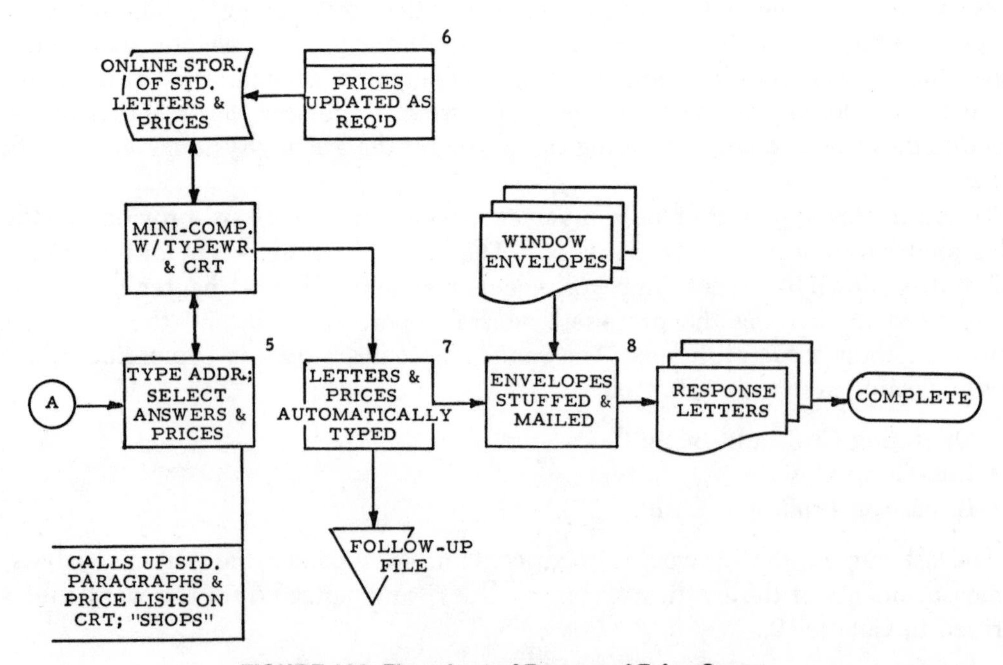

FIGURE 106: Flowchart of Proposed Price Query Answering System

PROPOSED-SYSTEM COST BREAKDOWN BY PROCESS STEP

SYSTEM Processing of Mailed-In Price Queries

Seq	Process	✓*	Direct Costs Per Day: Payroll	Equipment	Supplies	Other	Total
1.	Letters opened & sorted. Orders or Queries?		$ 5.60				$ 5.60
2.	Orders fulfillment process (separate procedure)						
3.	Deliver queries to Sales Quotation Department.		2.80				2.80
4.	Queries read						
5.	Addresses typed & answered & price lists selected on minicomputer		37.20	19.23	10.00	6.50	72.93
6.	Letters and price lists automatically typed.						
7.	Price lists periodically updated (separate procedure).					15.00	15.00
8.	Envelopes stuffed, sealed, stamped and mailed		11.20	1.20	3.20	15.60	31.20
						Subtotal	127.53
						Plus Fringe at 12 % of Payroll	6.82
						Plus Indirect at 43 % of Payroll	24.42
						Total	158.77
						Average No. of Units Processed Per	160
						Cost Per Unit	0.99

(Items 5 and 6 are bracketed together; the costs on their row cover both steps.)

*Place check mark at process step that is the pacing item. Describe limitation(s): Workload remains at average of 160 queries per day which is handled by one employee working with minicomputer. System can accommodate 240 queries per day without increase in personnel or equipment

FIGURE 107: Example Use of Proposed-System Cost Breakdown by Process Step Work Sheet

five employees with five typewriters have been replaced by a single employee with a minicomputer system with stored paragraphs and price lists. The costs are summarized on the work sheet shown in Figure 107, Example Use of Proposed-System Cost Breakdown by Process Step work sheet. This system approach shows a unit cost factor of 99 cents as compared to the existing system's unit cost of $2.45.

DEFINING SYSTEM OPERATION TASKS AND RESOURCE REQUIREMENTS

As just illustrated, an analyst must first determine all of the elements of a system so as to arrive at an estimate of its operating costs. An all-encompassing example of such a breakdown is illustrated in Figure 108. At the first level, it shows the "life" of the system as consisting of three basic elements:

- Design System
- Implement System
- Operate System

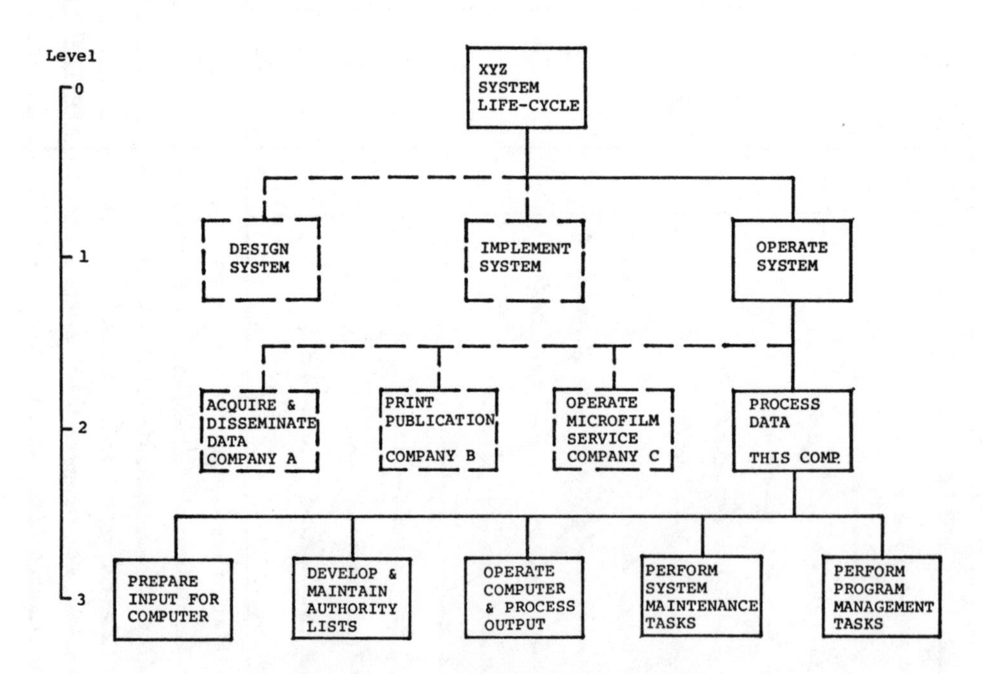

FIGURE 108: Example of Main Elements of a System's Life-Cycle

The first two are shown with broken lines to indicate that the costs of these parts of the system's life have been, or are being, determined elsewhere. In the case of "Design System," that cost was estimated at the beginning of the project and can now be confirmed or adjusted to meet actual costs. The cost of implementing the system is being determined as one of the important steps in the implementation planning process

(Chapter 17). Thus, the systems analyst is concerned, here, with defining and costing the tasks required for the third element, "Operate System." His chief source of information for constructing this breakdown is the New-System Flowchart or the New-System Flow work sheets.

In breaking down the "Operate System" to its next level of tasks, the example shows that one task ("Process Data") is performed by the company being studied, while three other tasks are sub-contracted to other companies. These latter three tasks, then, are shown as broken-line functions, the analyst being able to secure these costs in the form of vendor estimates or bids. This leaves the analyst with the job of defining only the "Process Data" task to greater levels of detail for purposes of establishing costs.

OPERATING TASKS BREAKDOWN **6.8**

Task No.	1	2	3	4	5	6	7	8	
	Operate System								
		Process Data							
100000			Prepare Input for Computer						
110000				Prepare Input from Department A					
111000					Prepare Data for Publication "A"				
111100						Perform Receiving Tasks			
111110							Enter Material Receipts on Log		
111120							Reconcile Material with Packing Slip		
111130							Check Original Document for Legibility		
111140							Make Duplication Check		
111141								Check Against Work Card File	
111142								Prepare New Card	
									Enter Author's Name
									Enter Report Number
									Enter Source Location
111150							Check for "Permission to Repro" Sticker		
111160							Shelve in Controlled Vault		
111170							Select for Processing Priority		
111200						Perform Descriptive Cataloging Tasks			
111210							Assign Accession Number		
111211								Enter Information on Accession Log	
111212								Affix Accession No. on Original Document	
111213								Affix Accession No. on Work Sheet	
111214								Affix Accession No. on Work Card	
111220							Enter Resume Date		
111230							Assign Program Area Code		
111240							Enter Document Title		
111250							Enter Author(s) Name(s)		
111260							Enter Instit. Names, Codes, Numbers		
111261								Enter Against Instit. Source Code List	
111262								If New, Prepare New Entry for SCL	
111270							Enter Contract No. if Applic. & Avail.		
111280							Enter Publication Date & Page Count		
111290							Calculate & Enter Microfilm Price		
111300						Perform Abstracting Tasks			
111310							Read Document		
111320							Write Abstract (Approx. 200 Words)		
111330							Edit, and Correct Rough Drafts		
111400						Perform Indexing Tasks			
111410							Select and Enter Descriptors		
111420							Fill Out Form #10 for New Descriptors		
111421								Enter Accession No. of Mat. Involved	
111422								Check All Available Authority Lists	
111423								Totally Define Descriptors	
111424								Indicate Relationship to Other Terms	

FIGURE 109: Example of the Use of the Operating Tasks Breakdown Work Sheet

RESOURCE ESTIMATING **6.9**

System Operations Tasks

Tasks / Resources: No.	Title	Manhours By Category				No. Iterations	Matls & Supplies	Equipt. Usage	Other

FIGURE 110: Resource Estimating Work Sheet —Systems Operations Tasks

The general rule for defining tasks, set forth in Chapter 2, is that they should be defined to a level where they can be easily managed and costs accurately determined. In the example shown (Figure 108), each of the five third-level tasks are too broad to accurately determine costs. The analyst, then, must define them to still further levels of detail. Figure 109 shows this done on an indentured list type breakdown, using the Operating Tasks Breakdown work sheet. If the analyst determined, as on this example, that tasks were defined to too fine a level of detail, he could make his estimates on the basis of higher level tasks.

The work sheet for estimating resources required for the system operations tasks is shown in Figure 110. At least the task numbers from the breakdown work sheet should be transferred to this one, which provides additional spaces for estimating man-hours by wage category, and recording the estimated requirements for other resources as well.

In the case of repetitive tasks (all of those on the example are repetitive), the analyst must factor his estimates by the estimated number of repetitions expected during the unit of time he has selected for purposes of comparing these operating costs with the previously defined existing-system costs.

Once completed and translated to cost figures, this information is summarized on the Operating Cost Comparison work sheet described next.

OPERATING COSTS AND BENEFITS

The first of four cost and benefit comparison work sheets is shown in Figure 111, the Operating Cost Comparison work sheet. It calls for a comparison of operating costs of the proposed system versus the existing system in two ways. First is a comparison on the basis of total system operating costs. Second, operating costs can be compared on the basis of either individual tasks, or in terms of outputs produced. In this latter case the current cost of producing a particular management report, for instance, can be compared with the proposed system cost.

As mentioned, at the point in the analysis where the analyst is ready to enter the estimated operating costs of the new system, this work sheet should already contain existing-system costs, derived from the work sheets described in Chapter 9. These costs should be broken down, first, in terms of direct labor and indirect charges. The rate factor used for indirect charges should be noted, and a definition of what costs the indirect charges cover should be entered at the bottom of the work sheet. Facility usage charges, for instance, might be included as part of the indirect charge. If not, then the analyst would have to provide a cost for that factor, itemizing it separately in one of the "Other" spaces.

The other major operating costs are also itemized on the work sheet. These can include computer usage, other equipment usage, the cost of directly chargeable supplies and materials, and the original costs of designing and implementing the systems, amortized over some agreed-to period of years. The analyst can use his judgment in separately itemizing any other type of significant operating cost.

General and Administrative expenses are entered on the work sheet next. This factor is usually a percentage of the sub-total costs.

When all costs for both the existing and proposed systems have been entered, the net gain or loss resulting from the new design is entered in the right hand column. If the proposed system shows significant cost advantages over the existing system, then there is less emphasis on the value of the other benefits derived from the system. On the other

6.10

OPERATING COST COMPARISON

System Operating Costs Per(Time)________

	Existing System	New System	Net Gain (Loss)
Direct Labor			
Indirect Costs (Rate________)*			
Sub Total			
Computer Usage			
Other Equipment Usage			
Direct Supplies/Materials			
Design & Implement'n Costs (Amortized)			
Other:			
Other:			
Other:			
Other:			
Sub-Total			
General & Administrative (Rate____)			
Total			

Unit Costs-Summarized by Tasks or Output

Unit:			
Unit:			
Unit:			
Unit:			
Unit:			
Unit:			
Unit:			
Unit:			

*Indirect Costs Defined________

FIGURE 111: Operating Cost Comparison Work Sheet

hand, if the proposed system costs more to operate than the existing system, then the justification for its adoption rests heavily on value improvements.

Values are compared on the Cost/Benefit Comparison Summary work sheet shown in Figure 112. There is no way to predict what would be significant value measurements without knowledge of the system being studied. The analyst, once again, must use judgment in selecting significant measurements from the Existing-System Value Measurements work sheets for entering on this work sheet and comparing with the same factors for the proposed system.

COST/BENEFIT COMPARISON SUMMARY **6.11**

	Existing System	New System
Operating Cost Per_____		
Economy (Other)		
Efficiency		
Productivity		
Quality		
Usability		
Accuracy		
Timeliness		
Regulations		
Reliability		
Adaptability		

FIGURE 112: Cost/Benefit Comparison Summary Work Sheet

This work sheet can also be supplemented with graphic charts and tables, if these aid in better visualizing the differences between the two systems.

LIFE-CYCLE COST COMPARISON

It is not a new idea that the important cost factor of any item is not necessarily its acquisition cost, but rather its "life-cycle" cost: the total cost of acquiring and operating it for some specified, logical period of time. An item that is the most expensive to acquire,

6.12

LIFE-CYCLE COST COMPARISON*

	1st Year	2nd Year	3rd Year	4th Year	5th Year
EXISTING SYSTEM-					
Direct Labor					
Indirect Costs (Rate____)					
Computer Usage					
Other Equipt Usage					
Direct Supplies/Materials					
Design Costs (Amortized)					
Implement'n Costs (Amortized)					
Other:					
Other:					
General & Admin. (Rate____)					
Total					
NEW SYSTEM-					
Direct Labor					
Indirect Costs (Rate____)					
Computer Usage					
Other Equipt Usage					
Direct Supplies/Materials					
Design Costs (Amortized)					
Implement'n Costs (Amortized)					
Other:					
Other:					
General & Admin. (Rate____)					
Total					

*Adjusted to reflect anticipated growth per Plans and Trends work sheets.

FIGURE 113: Life-Cycle Cost Comparison Work Sheet

for instance, might be the cheapest to operate. Thus, in the long run, it may be the most economical choice among several alternatives.

The total life-cycle cost of a system is of vital importance, and of special interest to the people who are buying the system. Once the new system has been designed, the systems analyst must determine this by examining and costing every facet of the system's life, from original concept through an operational period of time, the length of which is agreed to by management. For a data processing system, keeping in mind the continual evolution of new equipment and the need for keeping pace with competition, this might be a five-year period.

IMPACT ON PROFIT AND CASH **6.13**

	1st Year	2nd Year	3rd Year	4th Year	5th Year
EXISTING SYSTEM					
Sales Revenue					
Cost of Revenues					
Gross Income					
Expenses (less Syst. Op. Costs)					
System Operating Costs					
Income Before Taxes					
Taxes					
Net Income					
% to Sales					
Cash Balance - Beginning					
Cash Received (Projected)					
Cash Disbursed (Projected)					
Cash Balance - Ending					
NEW SYSTEM					
Sales Revenue					
Cost of Revenues					
Gross Income					
Expenses (less Syst. Op. Costs)					
System Operating Costs					
Income Before Taxes					
Taxes					
Net Income					
% to Sales					
Cash Balance - Beginning					
Cash Received (Projected)					
Cash Disbursed (Projected)					
Cash Balance - Ending					

NOTES: ______

FIGURE 114: Impact on Profit and Cash Work Sheet

The Life-Cycle Cost Comparison work sheet shown in Figure 113 provides space for a five year system "life." The figures to be entered here consist of the costs summarized on the Operating Cost Comparison sheet, translated to an annual cost basis (if they are not already in that form), and adjusted according to forecasts itemized on the Plans and Trends work sheet. These costs include, of course, the system design and implementation costs, as well.

In projecting future workloads, it would be useful for the analyst to show these on a minimum/maximum basis. A well-designed data processing system should appear very favorable, cost-wise, under maximum workload conditions.

IMPACT ON PROFIT AND CASH

If the system being studied is significant enough in scope in relationship to other company activities, it could have a measurable impact on company profit and cash. Furthermore, an electronic data processing system often involves a significant investment on the part of the company, and the resultant system and hardware represent an important asset. Under these circumstances it would be well to compare existing system and proposed system costs as they affect profit and cash. The work sheet shown in Figure 114 has been provided for that purpose.

In using this work sheet, the company's financial data is first recorded, subtracting the existing system's projected costs from total expenses and listing them separately. If the existing system costs are such that they are a portion of cost of sales, that figure would have to be adjusted accordingly.

Finally, in displaying the proposed system's figures, these projected costs are substituted for those of the existing system. If other adjustments must be made, an explanatory notation should be added at the bottom of the work sheet.

16
Determining Personnel/Training Requirements

Any change in a system implies a requirement for at least new knowledge and usually new skills on the part of operators, administrators, users, and managers. This in turn generates a requirement for orientation and either training or the acquisition of new personnel. With changing skill and knowledge requirements, organizational changes may also be required. An obvious example of this type of requirement would be when a system was being automated for the first time and major equipment being acquired.

These requirements will affect the feasibility analysis and, of course, the implementation costs. They may also affect operating costs if there are unique aspects to the new system and if a high turnover or rotation of personnel is anticipated.

Although requirements for training often seem highly intangible and nebulous and not susceptible to precise analysis, the approach to determining their effects is similar to analyzing other changes required. A baseline is established as to what exists. This is compared to what would be needed, the difference becoming the requirement. In this case the baseline and the new-system requirements are expressed in terms of skills. To this is added the amount of normal orientation that is needed for managers and other personnel.

A subtle distinction must be made between orientation and training. Orientation is simply the imparting of knowledge about the new system. This will be required, to varying degrees, for everyone in the organization affected by the new system. For some, orientation may require only a one- or two-page memo; for others, several hours of briefings. Training, on the other hand, has as its objective the imparting of new skills. It will be required by any whose tasks have been changed or for whom new tasks have been created. It will require such techniques as formal classroom training sessions, practice sessions, "programmed" training aids, monitoring of performance, and assistance on the job. The alternative to training is acquisition of new personnel who already possess the required skills. This will require an accurate and realistic set of position descriptions with specifications of skills required.

Sources for orientation and training will vary. In all cases the team of analysts will be required to perform a portion of this effort even if it is only to explain the system to professional trainers. The team may, of course, do all of the training and orientation as well as write the position descriptions if they are required. If the organization acquiring the new system is large enough to have a full-time training department, it may be utilized and the personnel department may be called on for assistance with position descriptions. One source of training assistance, if any equipment or off-the-shelf software packages are being acquired, will be the vendor from whom they are purchased or leased. The limitations relate to the emphasis of vendor training on his own product and the motivation, often unconscious but always present, to sell additional products. This only means that vendor training must be supplemented with additional system training or orientation. Some of the larger manufacturers also offer courses that are not specifically product-oriented. These could be used to advantage and are nominal in cost, if not free.

ORIENTATION

Orientation regarding the new system usually covers general areas. These include a general description of the purpose and reasons for system changes, how the new system works, and what changes to expect that affect the overall organization. In presenting this material, the first group to be considered is management. Since much of this type of orientation will have taken place through progress reviews, a summary type briefing is probably all that is necessary.

There will be other managers affected, however, probably a subordinate level to those receiving the progress reviews, and perhaps managers and supervisors from interfacing organizations. For this level, groups should be limited in size to between five and ten people and at the most twenty. This may mean several repeat sessions. It usually requires about eight hours of preparatory effort to get ready for a one or two hour session. To this must be added the cost of preparing training aids such as briefing charts or handouts at from one-half hour to four hours per chart or page. These sessions should generally be kept to one hour each, and an attempt should be made to keep them under four hours in length.

Topics that should be planned for these briefings will include a general overview of how the system works, what to expect in terms of management aids and reports coming out of the system (particularly in terms of changes from existing reports), what to expect in changes in operating costs, and what new requirements will be imposed on personnel that these specific managers and supervisors are responsible for. Depending on the management level, the new system's influence on profit may also be included. Training

ORIENTATION REQUIREMENTS **8.8**

Orientation Title/Purpose ____________________

Type of Orientation ____________________

Topics to be Covered ____________________

Orientation Length __________ No. Persons Requiring Orientation __________
No. of Separate Orientation Sessions __________ Avg Size, Each Group __________
If Briefing Form Used, No. & Background of Speakers ____________________

Tentative Orientation Schedule: ____________________

Orientation Site ____________________

Describe Travel or Subsistence Required, If any ____________________

Orientation Materials

Name	Quantity	Develop or Acquire	Cost

FIGURE 115: Orientation Requirements Work Sheet

and orientation planned for their personnel should also be explained, with emphasis on the reason for the training and how it will affect their specific departments, and its importance to the success of the system. In other words, the briefer must clearly explain the values of the new system so as to justify, to these managers, the "non-productive" man-hours to be spent by their personnel in the training sessions.

As to non-management personnel, everyone even remotely affected by the changes imposed by the new system should be given at least a brief indoctrination or notice regarding the new system. As mentioned earlier, a memo, preferably from the manager of the department concerned, may suffice. For others whose jobs are more directly affected, some additional orientation will be required. In planning for this task, the analyst can estimate, from his knowledge of the organization and the extent of the changes, which persons will require this orientation.

To aid in defining orientation requirements, an Orientation Requirements work sheet, shown in Figure 115, is provided. The top of the form provides space for identifying the title and/or purpose of the orientation, followed by a definition of its type such as briefing, memo, or notice.

The topics to be covered should be recorded next, then information related to the orientation's planned length, and the total number of persons requiring orientation should be noted. If the orientation is to be in the form of a briefing, the analyst should record the number of separate orientation sessions planned, and each session's average size in terms of attendees. Also, the number and background of speakers who will present the briefing should be noted.

The work sheet next provides space for constructing a tentative orientation schedule. The location should then be specified, and if travel or subsistence is required for either the attendees or the speaker, or speakers, that fact should also be recorded along with an estimate of the cost.

If orientation materials such as briefing charts are required, such materials should be identified on the work sheet. With this should be an estimate of the quantity required, a notation as to whether the material will be developed by the analyst or be acquired, followed by an estimate of what these costs will be.

TRAINING

It is predictable that the successful operation and maintenance of a new system will depend heavily on the acquisition, through hiring or training, of very specific skills. Three separate work sheets are provided to help in this definition effort.

The Skills Inventory Work Sheet

The first step in analyzing training requirements is to make an inventory of skills already available. The work sheet shown in Figure 116 is designed to facilitate this inventory. Depending on the size and complexity of the organization being studied, a separate work sheet may be made out for each sub-unit, or one for the total organization may suffice.

The identification of the organization is entered in the space provided at the top of the form and the principal tasks or functions performed listed down the column on the left. As each task is recorded, the skills and specialized knowledge required to perform that task are recorded in the spaces to the right. The experience level is also noted, usually in terms of years of experience or, in some cases, a rate of production. At the far right is a column for entering the number of personnel currently available in the organization that have the skill listed.

An example of a type of entry that would be made on the work sheet might be:

8.9

SKILLS INVENTORY

Organization Unit ______________________

Tasks or Functions	Skills Used	Experience Level	No. of Persons

FIGURE 116: Skills Inventory Work Sheet

- (Task) "Keypunch billing data"
- (Skills Used) "Operate model XX keypunch"
- (Exp. Level) "One Year" (Or this could be in terms of characters or cards per minute.)
- (No. of Persons) "One"

The Skills Requirements Work Sheet

The next step is to analyze the skills required for the new system, then matching these requirements with what's available so that the extent of hiring or training can be

8.10

SKILLS REQUIREMENTS

Task No.	Skills Required	No. Req'd	No. Avail.	Net Diff.	Hire	Train

FIGURE 117: Skills Requirements Work Sheet

determined. This is done on the Skills Requirements work sheet shown in Figure 117. Tasks are recorded in terms of their identifying code number. The significant skills that will be required to perform each task are listed in the column to the right of the task number. Once all of the tasks and corresponding skill requirements have been enumerated, this list is compared with the skills inventory. Where requirements can be matched to existing skills, they are so noted. At the same time the "Net Difference" column is filled out, indicating any difference between the skill required and the corresponding skill that is available.

Finally, the analyst records a recommendation as to whether the difference should

TRAINING REQUIREMENTS 8.11

Training Course Title ____________________

Type of Training ____________________

Topics to be Covered ____________________

Course Length (Hours) __________ Number of Trainees __________
No. of Separate Training Groups __________ Avg. Size, Each Group __________
No. & Type of Instructors ____________________
Tentative Class Schedule: ____________________

Classroom or Training Site ____________________

Describe Travel or Subsistence Requirement, If Any ____________________

Training Materials:

Name	Quantity	Devel. or Acquire	Cost

FIGURE 118: Training Requirements Work Sheet

be made up by training existing personnel, or by acquiring new personnel, or a combination of the two.

Training Requirements Work Sheet

Once skill requirements have been defined, the next step is to determine the actual training requirements. This is done by reviewing the Skills Requirements work sheet and grouping similar requirements into logical course-of-study categories. The requirements for each category of training are developed on separate Training Requirements work sheets, illustrated in Figure 118. The course title is first specified, followed by a definition of the type or method of training that is planned. This might include classroom instruction, or on-the-job training, or outside school or training center, or vendor's training course, or other similar methods.

The systems analyst then records the main topics that should be covered in this particular course. This is followed by notations regarding the course length, the total number of trainees, the number of separate training groups, and the planned average number of trainees in each group. The number and type of instructors required for this particular course of study is also specified.

The analyst can then record a tentative class schedule on the work sheet, describe the planned location, or locations, of the classes, then describe any travel or subsistence factors that may apply.

The bottom part of the work sheet is to be used for defining the training materials (such as text books, visual aids, demonstration equipment, etc.) that are required for this course of study. The quantity needed is also noted, together with the information as to whether this material will be developed or acquired, and the estimated costs involved.

The information recorded on the work sheets described in this chapter provides one of the essential inputs to the planning and pricing of the new system's implementation. This subject is covered in the following chapter.

17

Using the Workbook to Plan System Implementation

Once the new system has been designed, it's the systems analyst's job to develop the plan for its implementation. What will be submitted to management for approval at the completion of this activity, then, is a two-part report:

- A Description of the Proposed New System (including cost)
- The Plan for the New System's Implementation

More often than not, the management personnel charged with the responsibility of overseeing the development of new systems are not familiar with the details of data processing systems. For this reason, it is also the responsibility of the systems analyst to see that both the system description and implementation planning materials are prepared in such a manner so they can be easily understood and evaluated. Previous chapters described how this can be done with the description of the new system. Those chapters showed, for instance, how requirements and preferences should be cross-referenced to the graphic flowchart so that compliance to these factors could be readily determined. It was also shown how costs and benefits should be summarized so management could easily examine the estimated value of the proposed system.

This chapter deals with the task of developing the implementation plan so it, too, can be readily understood and evaluated. The subject first covered is the development of project specifications. This is followed by a description of the steps involved in planning the project, and concludes with a summary of the materials available for submittal to management for review and approval.

PROJECT SPECIFICATIONS

At the beginning of the systems analysis project, it will be remembered, two types of specifications were prepared: the System Specification and the Project Specification. Once the new system has been "specified" in terms of flowcharts, descriptions, equipment configuration, and other such materials, it remains only to specify the project requirements prior to detailed planning.

Implementation Project Specification Work Sheet—System Development Planning Factors

The Workbook, in its Implementation Plan section, provides four project specification work sheets. The first, the Implementation Project Specification—System Development Planning Factors, is shown in Figure 119. It lists major categories of tasks that might be required in the implementation phase in order to translate the proposed new system into an operating system. It is much like a checklist, but it also calls for the systems analyst to make brief descriptive notations that clarify tasks to be performed, and to roughly estimate time factors involved in their accomplishment. The time factors are merely guidelines at this stage, for detailed estimates can be more easily determined later, after the tasks have been broken down to a lower level of detail.

The first entry on this work sheet involves a brief statement specifying the type of detailed specifications that may be required as a next step in system development. This might be the segmenting of process runs, for instance. Spaces are next provided for briefly stating task requirements relating to the developing of program flowcharts and decision tables, and for the coding and checking of programs.

Potential implementation phase tasks relative to tests are broken into three categories on the work sheet. The first provides space for briefly stating task requirements relative to the development of material to be used in conducting tests. The second category involves the testing and debugging operations, followed by the integrated system test function.

Tasks involving the installation of the software on the customer-designated equipment, pilot operations, and the training of customer personnel are to be briefly described next on the work sheet. Finally, space is provided for briefly describing any other system development task that doesn't fit under the other categories listed on the work sheet. As with all the work sheets, supplemental sheets should be used where the space provided is not sufficient.

Implementation Project Specifications Work Sheet—Files/Documents/Documentation Planning Factors

The second of four Implementation Project Specification work sheets is shown in Figure 120. This work sheet provides spaces for briefly describing potential implementation phase tasks regarding manual and automated files, documents, and system documentation.

8.1

IMPLEMENTATION PROJECT SPECIFICATION

System Development Planning Factors

	Time Factors
Detailed Specifications	
Program Flowcharts & Decision Tables	
Descriptions	
Coding and Checking	
Test Data	
Testing and Debugging	
Integrated System Test	
Software Installation	
Pilot Operations	
Training Programs	
Other	

FIGURE 119: Implementation Project Specification Work Sheet—System Development Planning Factors

The first two entries are in regards to the drafting and reproduction of forms that will be used in the new system. A brief description of tasks involved in the development of new, or the conversion of existing, manual files and data bases is called for next. The subject of system documentation is covered next in four separate categories. The first involves the assembling of system analysis and design materials into an "Application Manual" so that all of the original study material, design specifications, and rationale can

8.2

IMPLEMENTATION PROJECT SPECIFICATION

Files/Documents/Documentation Planning Factors

	Time Factors
Forms Graphics	
Forms Reproduction	
File Conversion	
New Files	
Data Base Conversion	
New Data Base	
Application Manual	
Master Run Manual	
Operating Instructions	
Manual Procedures	
Other	

FIGURE 120: Implementation Project Specification Work Sheet—Files/Documents/Documentation Planning Factors

be made readily available in organized form for future reference. The second involves the tasks required for developing and assembling the various program development materials (program flowcharts, decision tables, descriptions, program listings, input/output formats, etc.) into an organized system of documentation. The third and fourth entries involve the development of machine operating instructions and manual procedures. (See Chapter 19.)

IMPLEMENTATION PROJECT SPECIFICATION 8.3

Equipment & Facility Planning Factors

	Time Factors
Detailed Site Layouts	
Site Preparation	
Equipment & Cable Ordering	
Supplies Ordering	
Equipment Delivery	
Equipment Installation	
Equipment Check-Out	
Other	

FIGURE 121: Implementation Project Specification Work Sheet—Equipment and Facility Planning Factors

Implementation Project Specification Work Sheet—Equipment and Facility Planning Factors

If the equipment configuration for the proposed system involves the acquisition of new equipment, tasks related to that acquisition must be determined. To aid in this preliminary determination, the Implementation Project Specification work sheet with equipment and facility planning factors is provided, and is illustrated in Figure 121.

8.4

IMPLEMENTATION PROJECT SPECIFICATIONS

Schedule and Location Considerations

Schedule Factors

Implementation-Start Considerations ______

Implementation-Complete Considerations ______

Implementation Area Geographical Location(s) ______

If Travel & Subsistence Required, Specify ______

Will Client Provide Work Space? ______

Specify ______

Can Customer Equipment Be Used For Testing and Debugging Tasks? ______

If So, Describe ______

FIGURE 122: Implementation Project Specification Work Sheet—Schedule and Location Considerations

Space is provided on this work sheet for briefly specifying the tasks involved in developing detailed site layouts, preparing the site, ordering equipment, cables, and supplies, delivery of the equipment, its installation, and its check-out. If other tasks are anticipated they, too, can be specified.

Implementation Project Specification Work Sheet— Schedule and Location Considerations

The fourth Implementation Project Specification work sheet is shown in Figure 122. It is used for specifying schedule and location considerations that will serve as vital input to the planning process. It is virtually identical to one of the project specification work sheets used to originally plan the analysis and design portions of the overall project.

Factors governing the start and completion of the implementation should be first specified. There could be restraints, for instance, imposed by difficulties in securing satisfactory delivery dates for equipment. Possibly, implementation of the new system would best serve the user by being accomplished in "stages." Notations to this effect should be entered on the work sheet.

Entries should next be entered regarding geographical locations together with travel and subsistence requirements, if any. If work space can not be provided by the client, and if the client's equipment is not available or satisfactory for testing and debugging operations, these factors must be also noted so that other arrangements can be planned.

PROJECT PLANNING

Once the Implementation Project Specification work sheets have been completed, those materials, together with other Workbook materials relating to the proposed system design, serve as the input for the development of the implementation plan. As with all planning, the first step involves defining the detailed tasks required to achieve the specified objective.

The Implementation Task Planning Work Sheet

The Implementation Plan section of the Workbook provides a task planning work sheet (Figure 123) identical, except for sub-title, to the task planning work sheet used to originally define the analysis and design portions of the project.

On this "indented" list, tasks are broken down, control numbers assigned, and outputs are specified where possible. The major categories of tasks that might be required to implement a system are shown in graphic form in Figure 124, Example of Breakdown of Tasks for Implementation of System.

How a breakdown of tasks is structured is a matter that the systems analyst must decide on the basis of the project being defined. It should be done so that the tasks are easily understood, each level further defining the level above it in the structure. As with all such planning, the tasks should be broken down to a level where they can later be readily and accurately priced. It is also important for the analyst to call in the key people who will be responsible for performing the work so they can define their own tasks.

After the detailed tasks required for implementing the system have been defined and translated into a narrative "Work Statement," they must be scheduled. The first step in this process is determining the interrelationships of the tasks by constructing a network such as described earlier in Chapter 2. Time factors are then assigned, restraints and "critical paths" are identified, adjustments in the allocation of resources made, and a final project schedule developed. This latter activity must be done in parallel with the defining of resource requirements.

A consideration at this juncture is whether there should be an instant or gradual changeover to the new system. Both concepts are illustrated in Figure 125. A point to

8.5

TASK PLANNING
Implementation Plan

Control No.	Level						Output
	0	1	2	3	4		

FIGURE 123: Implementation Task Planning Work Sheet

consider regarding the gradual change is the added cost of operating parallel systems until the changeover is completed.

Implementation Resource Estimating Work Sheet

Manpower and other resources required to accomplish the defined implementation tasks are developed on the Implementation Resource Estimating work sheet shown in Figure 126. The format and use of the work sheet is identical to that described earlier in Chapter 2. In its application, however, the determination of service-type requirements

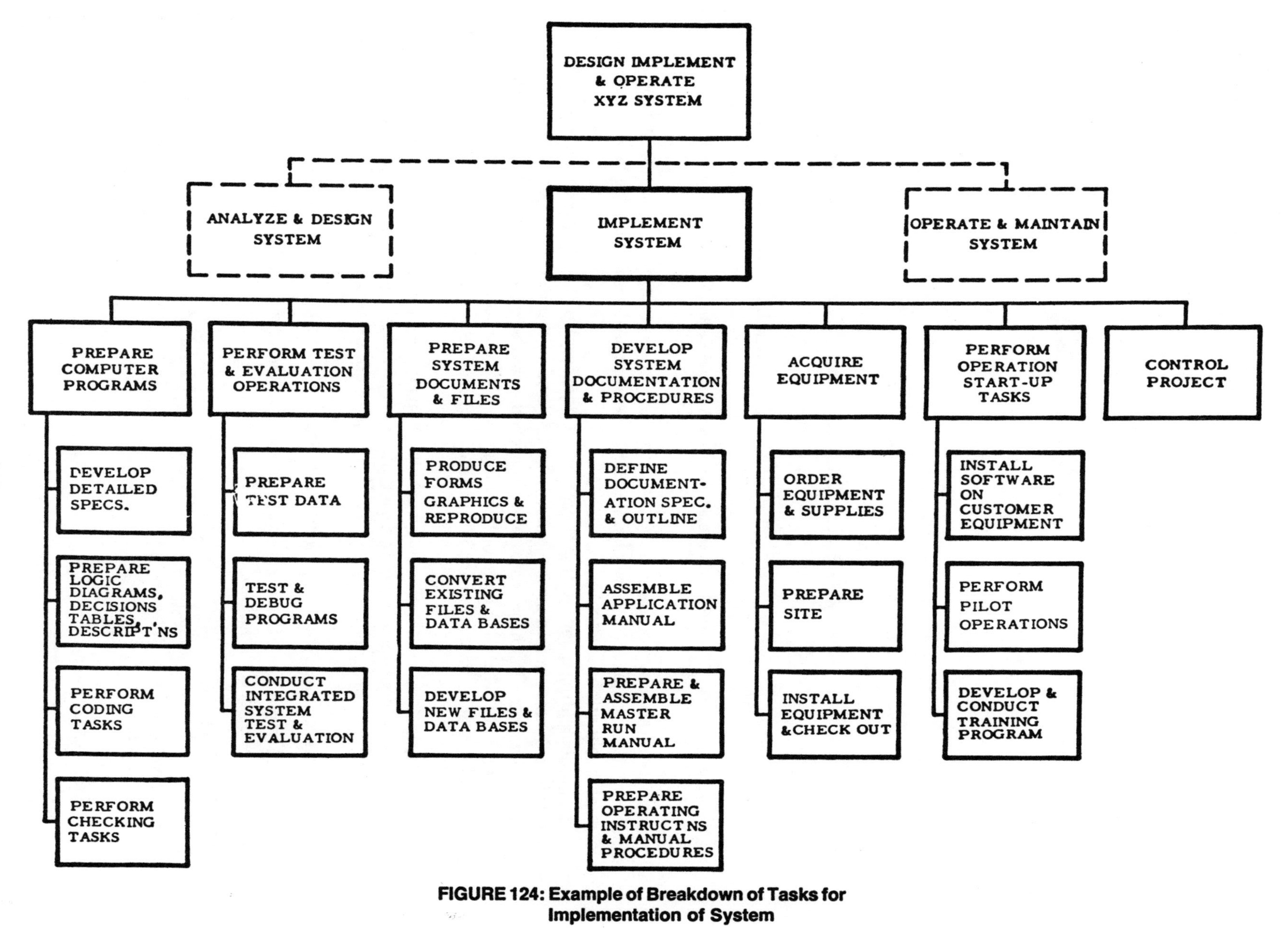

FIGURE 124: Example of Breakdown of Tasks for Implementation of System

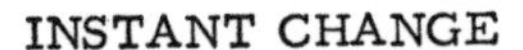

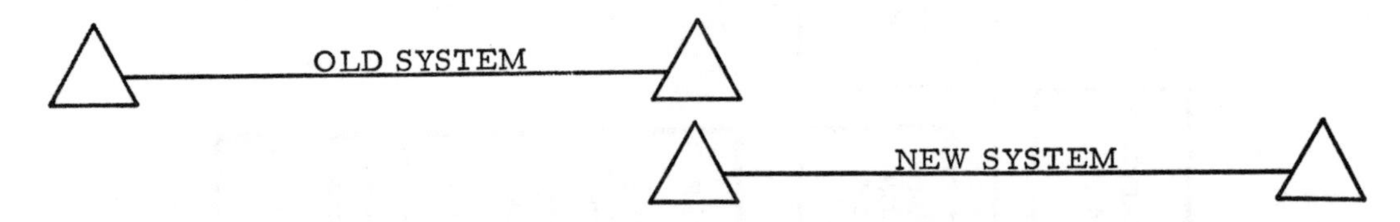

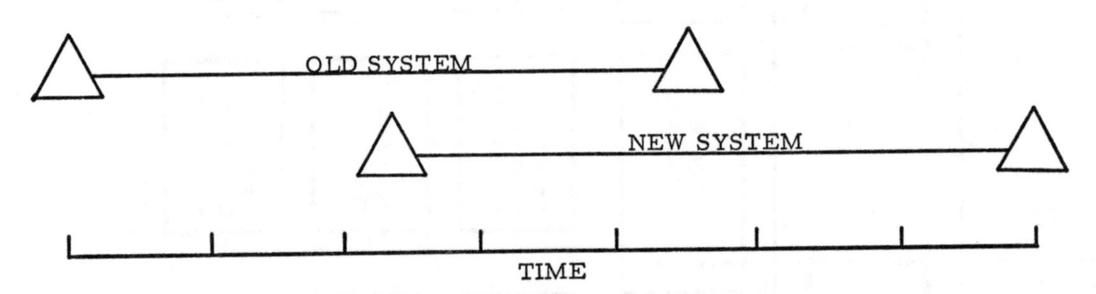

FIGURE 125: Instant Versus Gradual Change

(reproduction services, computer services, non main-frame computer services), will be significant in this case, whereas in planning for the earlier analysis and design tasks these were relatively minor factors.

In implementing a new system there can be significant costs other than manpower and services. If new equipment is required, for instance, or if certain portions of the work are to be subcontracted, these costs must be secured from the appropriate bids and added to the other costs. The advantage of breaking down *all* the work required to achieve a stated goal is that, in this approach, it is difficult to inadvertently overlook even minor work and cost factors.

In using the Resource Estimating work sheet, manpower estimates can be identified in terms of individuals (and their corresponding wage scales) or in terms of skill types (systems analysts, programmers, coders, machine operators, etc.).

Implementation Cost/Price Summary Work Sheet

Once all implementation costs have been determined they can be summarized into major categories such as follows:

- Direct Labor
- Indirect Costs (sometimes a standard rate per hour of direct labor)
- Travel and Subsistence
- Equipment
- Supplies
- Site Preparation
- Computer Services

8.6

RESOURCE ESTIMATING

Implementation Plan

Tasks / Resources		Manpower (Indiv. or Skill)				Repro.	Comp. Serv.	Non Main-frame Comp.
No.	Title							

FIGURE 126: Implementation Resource Estimating Work Sheet

- Non-Main-Frame Computer Services
- Reproduction Services
- General & Administrative Costs (usually a standard rate of other costs)
- Fee

To aid in this summarization, the Implementation Cost/Price Summary work sheet is provided. It is illustrated in Figure 127.

IMPLEMENTATION COST/PRICE SUMMARY 8.7

	Rate	Hours	Cost/Price
Direct Labor:			
DL Category			
DL Category			
DL Category			
DL Category			
Indirect Costs			
Sub-Total			
Travel & Subsistence			
Equipment			
Supplies			
Site Preparation			
Computer Services			
Non Main-Frame Computer Services			
Reproduction Services			
Other			
Other			
Sub-Total			
General & Administrative Costs			
Sub-Total			
Fee			
Grand Total			

FIGURE 127: Implementation Cost/Price Summary Work Sheet

Technical Performance Review Materials Checklist

The plan for achieving technical excellence on a project is one of the most difficult to construct and one of the easiest to overlook. In any type of project it is always a formidable task to develop the plan for achieving quality, accuracy, reliability and other performance criteria. It is especially difficult to develop such a plan for controlling the development of computer software. There are, however, several basic techniques that can be utilized. The first and most obvious is the use of various tests and demonstrations to meet the criteria established in the system design. The second is the establishment of a system of

TECHNICAL PERFORMANCE REVIEW MATERIALS **8.12**

Checklist

	Check
Detailed Specifications	
Logic Diagrams	
Decision Tables	
Descriptions (Programs, Process, Run, etc.)	
Computer System Documentation Standards	
Computer System Documentation Outline	
Assembled Computer System Documentation	
Operating Procedures	
Operating Instructions	
Input Formats	
Output Formats	
Other System Documents Formats	
Forms Graphics	
Reproduced Forms	
Inital System Outputs	
Training Curriculums	
Training Schedules	
Training Aids	
Training Manuals	
Subsystem Tests	
Integrated System Acceptance Test	

FIGURE 128: Technical Performance Review Materials Checklist

progressive review and approval (by the project manager and the customer) of the evolving forms, manuals and other system documentation. Finally, there are Design Reviews that can be scheduled to establish formal control over the developing design. For this latter purpose, Design Review Checklists, such as described in Chapter 2, can be developed and used.

The development of the technical performance plan should occur early in the planning process, for its development often results in the creation of new tasks that should be included in the breakdown of tasks.

In developing technical performance criteria the category that comes first to mind is

"Tests and Demonstrations." The design for the system should specify certain parameters of technical performance. These might be related to quality, speed, accuracy or other measures of technical effectiveness that should be achieved in the total system operation. Tests and demonstrations for these parameters can be established and scheduled as an important part of the total project. These tests can be in several categories, one being subsystem tests as each is either completed or modified. The final test, of course, would be the integrated system test. In nearly every case these tests require the development of either real or simulated input test data and, in some cases, data-base and authority-list data. The development of this material should not be overlooked as a significant task in the previously described definition of tasks.

Throughout the implementation phase of a project a number of reports, forms and system documentation items are produced. The development of each one of these items can be broken up into at least several definable steps. The design of input forms, for instance, can be divided into:

- Data Element Analysis;
- Rough Draft Format Design;
- Test of New Form (under actual working conditions with some measurement noted of their effectiveness over the older system);
- Development of Final Form.

A management review and approval cycle of each of these steps can be programmed into the plan. In this manner management is able to progressively review and approve separate elements of the system as it evolves from a qualitative standpoint. A similar approach can be taken with the other materials developed during system implementation. Figure 128 displays a checklist of technical performance review materials that could be subjected to this type of review and approval.

Assigning the Tasks

The scope and costs of implementing a system usually far exceed that of analyzing the existing system and designing a new one. One of the ways this is reflected is in the greater number of people and broader range of skills required in the implementation of a system. For this reason a more detailed level of planning the grouping and assignment of tasks is called for.

The first step is to logically group the tasks according to the needs of the project and the types of skills required and available. This grouping can be graphically displayed as shown in Figure 129. In this approach only the top number in a given cluster (tree) of tasks need be shown. In the illustration, for example, the number 11000 appears under "Systems Analysis." This would indicate that if there were other tasks starting with the same number (11100, 11200, 11210, etc.) they would also fall under the responsibility of the Systems Analysis unit since these numbers appear nowhere else in the structure. A variation of this chart is where task titles are listed in addition to task numbers.

If there are a great number of tasks to perform during implementation, it might be well to develop a matrix of tasks versus assignments such as shown in Figure 130. A Task

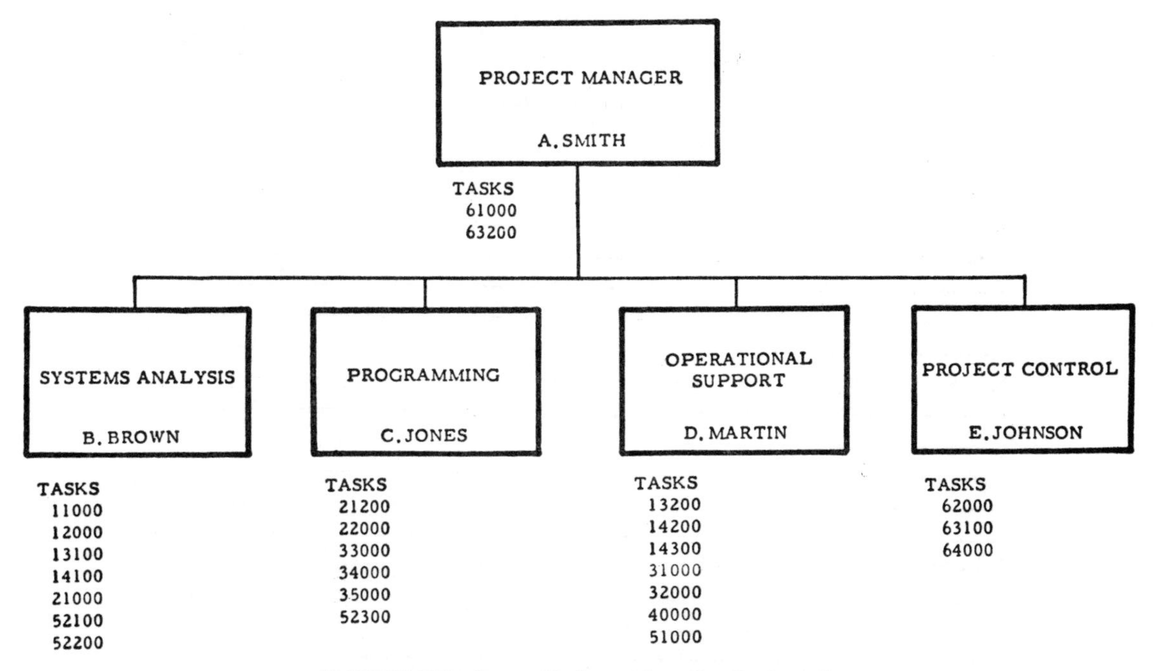

FIGURE 129: Cross-Referencing the Project Organization Chart with the Tasks

TASK ASSIGNMENT MATRIX **8.13**

Task No.	Task Title / Assignment	JHL	RPL	JAJ	LTN	PNS	TLT	ROS	JPM
1000	PREPARE COMPUTER PROG'MS	X LEAD							
1100	DEV'P DETAILED SPECS				X				
1120	PREPARE LOGIC DIAGRAMS			X			X		
1300	PREFORM CODING TASKS		X			X			X
1400	PERFORM BASIC CHECKOUTS		X			X			X
2000	PERFORM TEST & EVAL.	X LEAD							
2100	PREPARE TEST DATA							X	
2200	CONDUCT SUB-SYST TESTS		X			X			X

FIGURE 130: Example of a Portion of a Task Assignment Matrix Work Sheet

Assignment Matrix work sheet is provided, for this purpose, in the Implementation Plan section of the Workbook. Such a matrix helps assure that all defined tasks for the project have been assigned.

8.14

TASK ASSIGNMENT

Implementation Plan

Task No. ____________

Task Title ____

Task Description ____

____

____

____

____

____

Specific Output(s) Required ____

____

____

____

____

____

The Start Of This Task Depends Upon ____

____

____

____

____

Subsequent Tasks Dependent Upon This One's Completion Are: ____

____

____

____

____

____

Schedule

Start Date ____

Completion Date ____

Resources:

Manhours ____

Materials/Services ____

Other ____

Assigned To ______________________________ Date ____________

Assigned By ______________________________ Date ____________

FIGURE 131: Implementation Plan Task Assignment Sheet

As with the initial project planning described in Chapter 2, there is also a Task Assignment work sheet provided for implementation tasks. It is shown in Figure 131. It is used for summarizing all the planning data relating to a single task for purposes of clearly communicating the assignment. In this way it can serve as a project "work order."

Setting Up Controls

When the plans are complete, the material for controlling those plans must be developed. As mentioned earlier in Chapter 2, this could consist of schedule control charts, cost control charts, and technical performance control charts and lists.

Regardless of the analyst's planning excellence, however, there is one overriding consideration. The credibility of the entire plan, and the systems analyst's ability to perform to that plan, will be judged by management's observance of the analyst's performance during the earlier analysis and design portions of the project.

18
Presenting the System Concept to Management

"Final Report" is the name often applied to the combination of proposed system design and implementation plan materials presented to management. This report is easy to assemble when the workbook approach to systems analysis has been used. Additional work is necessary, however, if management wishes to have the report augmented with an oral presentation.

FINAL REPORT CHECKLISTS

Three checklists that will help insure that the final report is complete with all supporting data required are shown in this section. They will also be useful in assembling data on which the oral presentation will be based.

The first of these, the System Design Material Checklist shown in Figure 132, is a checklist of *Workbook* work sheets and generated documents that describe the proposed system, its new documents and files, and program description data.

The checklist of Plan and Control Materials is shown in Figure 133. This lists the available materials that constitute what is normally referred to as the "management" portion of a final report.

The third is a checklist, apropos for consultants, of items that should be considered when developing the agreement for performing the implementation tasks. This is the Contract Considerations Checklist shown in Figure 134.

8.15

SYSTEM DESIGN MATERIAL

Checklist

	Check
SYSTEM DESCRIPTION	
System Synopsis	
Flowchart	
Statement of Requirements	
User Preference	
System Flow Narrative Description	
System Features Summary	
Organization Change Factors	
Equipment Configuration	
Site Preparation Factors	
Training Factors	
Manual Operating Procedure Requirements	
Cost/Benefit Summary	
DOCUMENTS AND FILES	
Input/Output Synopsis	
Document Mock-Ups	
File Descriptions	
Data Base Descriptions	
Data Element Analysis	
GENERAL PROGRAM DESCRIPTION	
Program Flowchart	
Decision Tables	
Descriptions (Programs, Processes, Runs, etc.)	

FIGURE 132: System Design Material Checklist

NEED FOR ORAL PRESENTATION

The final report is necessary to document the system design study and to provide all the data needed for those assigned to study and evaluate its conclusions and recommendations. In many cases, decision makers also expect a concise briefing when they are

8.16

PLAN AND CONTROL MATERIAL

Checklist

	Check
PLAN	
Project Summary	
Statement of Work	
Indented List of Tasks	
Work Breakdown Structure	
Project Network	
Trip Plan	
Resource Plan	
Technical Performance Plan	
Organization & Assignments	
Price/Cost	
CONTROL	
Control System Description	
Milestone Tracking Schedule	
Trip Tracking Schedule	
Budget Tracking Charts	
Manloading Control Chart	
Equipment/Facilities Utilization Control Charts	
Materials Utilization Control Chart	
Technical Performance Control Charts	
RELATED DATA	
Resumes of Key Personnel	
Synopsis of Experience Related to This Project	
Description of Equipment & Facilities to be Used	

FIGURE 133: Plan and Control Material Checklist

presented with the basic facts relative to the study. The oral presentation can be beneficial to all parties, allowing a direct exchange of ideas and reactions to ideas. It provides an environment where questions can be asked which elicit further explanation of points that may not have been clearly understood.

While the oral presentation at the conclusion of the system design and implementa-

8.17

CONTRACT CONSIDERATIONS

Checklist

	Check
Start Date	
Contract Duration	
Payment Amount	
Payment Schedule	
Disclaimer of Other Agreements	
Rights to Make Commitments	
Control of Proprietary Materials	
Access to Classified Materials	
Release or Resale of Design Materials	
Advertising or Publicity Conditions	
Use of Customer Equipment	
Use of Customer Facilities	
System Augmentation Options	
Maintenance Support	
Warranty	
Guarantee	
Error Liabilities	
Delay Liabilities	
Other Liabilities	
Deviation from Plan Penalties	
Deviation from Design Penalties	
Cancellation Conditions	

FIGURE 134: Contract Considerations Checklist

tion planning cycle is the most usual requirement, presentations may also be requested at other phases of the project. The work sheets described here can be used for these other purposes as well.

PRESENTATION CONTEXT

The logic of the presentation usually follows a pattern consisting of the following elements:

- The Problem (And its cost)
- The Solution (The proposed system)
- The Plan (How it is planned to implement the solution)
- The Benefits and Costs (Why the project should be continued)
- The Recommendation
- The Desired Management Action

Within that framework, the analyst will usually be comparing two or more courses of action. In some cases this involves presenting several alternative solutions to the same problem. As a minimum, the proposed new system will be presented in relationship to the existing system, thus offering as one alternative the option of doing nothing. In other words, not to implement a new or changed system. Whether actually presented or not, it must be recognized that this alternative will have a strong appeal to some members of the audience.

The analyst should keep in mind that the management presentation is a summary of what is in the *Workbook*. Summarizing this material properly is an essential task in preparing the presentation. A beginning point might be a management-oriented version of the proposed system's flowchart where key features have been "flagged" to point out their salient features. An example of a small portion of such a flowchart is displayed in Figure 135. (This particular example represents a photoprocessing vendor's description of a proposed system for handling a particular customer's overload and special work.)

Summarizing material from the *Workbook* depends on selecting the essential facts to present and deciding which areas will be emphasized. Selection occurs at two levels. First, the presenter must select the major items necessary to build the logic. This provides little more than an outline of the presentation. Then he must select the significant aspects of these main points which will convince the audience of their validity.

The first level of selection is the easiest. To illustrate, the six items in the typical presentation outline given above could be covered as follows:

> Our secretaries spend only 30 percent of their time typing and are idle another 20 percent of the time waiting for managers to review what they have typed. The solution to this problem is a centralized word processing center which would operate continuously and would provide typing support to all our managers. We could realize a net saving of $50,000 a year implementing such a center. Therefore, we ask that you authorize the creation of the center organization and allocate funds for its development and operation.

The next step is to take this basic outline and expand on it to demonstrate its validity.

In selecting the facts to present, the analyst must consider where the emphasis in his presentation should be. This depends on the audience. If the key decision maker is the manager who pointed out the problem in the first place and asked for the study, there is no need to emphasize the problem. On the other hand, if the decision maker is well removed from the problem, much more time may have to be used convincing him that a problem does exist before the other points will be believable.

For the manager providing technical support, the analyst should be prepared with plenty of back-up information as to the technical feasibility of the solution. On the other hand, for the chief executive the statement that the solution is technically feasible may be sufficient.

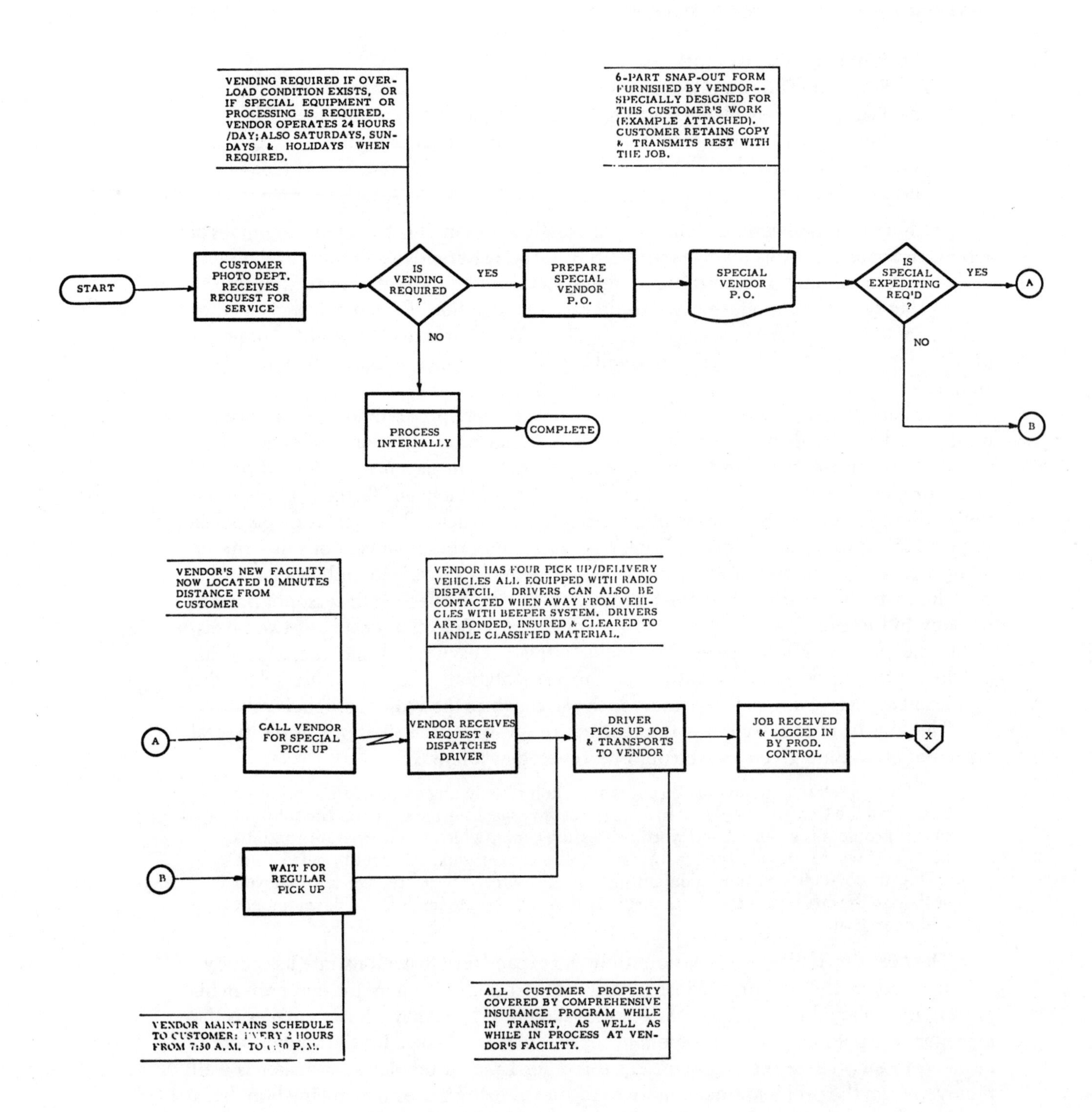
VENDING REQUIRED IF OVER-LOAD CONDITION EXISTS, OR IF SPECIAL EQUIPMENT OR PROCESSING IS REQUIRED. VENDOR OPERATES 24 HOURS /DAY; ALSO SATURDAYS, SUN-DAYS & HOLIDAYS WHEN REQUIRED.
6-PART SNAP-OUT FORM FURNISHED BY VENDOR-- SPECIALLY DESIGNED FOR THIS CUSTOMER'S WORK (EXAMPLE ATTACHED). CUSTOMER RETAINS COPY & TRANSMITS REST WITH THE JOB.
START
CUSTOMER PHOTO DEPT. RECEIVES REQUEST FOR SERVICE
IS VENDING REQUIRED ?
YES
NO
PREPARE SPECIAL VENDOR P.O.
SPECIAL VENDOR P.O.
IS SPECIAL EXPEDITING REQ'D ?
YES
NO
A
B
PROCESS INTERNALLY
COMPLETE
VENDOR'S NEW FACILITY NOW LOCATED 10 MINUTES DISTANCE FROM CUSTOMER
VENDOR HAS FOUR PICK UP/DELIVERY VEHICLES ALL EQUIPPED WITH RADIO DISPATCH. DRIVERS CAN ALSO BE CONTACTED WHEN AWAY FROM VEHI-CLES WITH BEEPER SYSTEM. DRIVERS ARE BONDED, INSURED & CLEARED TO HANDLE CLASSIFIED MATERIAL.
A
CALL VENDOR FOR SPECIAL PICK UP
VENDOR RECEIVES REQUEST & DISPATCHES DRIVER
DRIVER PICKS UP JOB & TRANSPORTS TO VENDOR
JOB RECEIVED & LOGGED IN BY PROD. CONTROL
X
B
WAIT FOR REGULAR PICK UP
ALL CUSTOMER PROPERTY COVERED BY COMPREHENSIVE INSURANCE PROGRAM WHILE IN TRANSIT, AS WELL AS WHILE IN PROCESS AT VEN-DOR'S FACILITY.
VENDOR MAINTAINS SCHEDULE TO CUSTOMER: EVERY 2 HOURS

FIGURE 135: Portion of Flowchart Used for Presentation Purposes

The analysis of the specific audience is one guide to summarizing. The other guide must be the objective of the presentation. Every communication has an objective to elicit a state of mind and resulting action on the part of the receiver or audience. The most efficient communications contain only those elements that are necessary to achieve this objective.

THE PRESENTATION CONSIDERATIONS WORK SHEET

The Presentation Considerations work sheet, Figure 136, is designed to help the

PRESENTATION CONSIDERATIONS **8.18**

ANTICIPATED PRESENTATIONS

	Go Ahead Needed For:				
Management Level	Concept				Imple-ment.
A. Immediate Supervisor					
B. Middle Management-User					
C. Mid. Manage. -Technical					
D. Mid. Manage. -Affected					
E. Upper Level Manager					
F. Executive Officer					

KEY PRESENTATIONS

1. To Whom ________ Approx. Date ________
   Action Desired ________ ____Workbook Review Only
   Major Criteria ________ ____Informal Meeting
   Back-Up Needed: ____Formal Briefing
   Work Sheets ________ Preferred Media
   Other ________ ________
2. To Whom ________ Approx. Date ________
   Action Desired ________ ____Workbook Review Only
   Major Criteria ________ ____Informal Meeting
   Back-Up Needed: ____Formal Briefing
   Work Sheets ________ Preferred Media
   Other ________ ________
3. To Whom ________ Approx. Date ________
   Action Desired ________ ____Workbook Review Only
   Major Criteria ________ ____Informal Meeting
   Back-Up Needed: ____Formal Briefing
   Work Sheets ________ Preferred Media
   Other ________ ________

FIGURE 136: Presentation Considerations Work Sheet

analyst anticipate and plan for the presentations to be given during the project. It should be completed early in the planning phase, along with other planning work sheets.

The matrix at the top of the checklist is primarily a scheduling aid. Across the top of the matrix are spaces for indicating the decision for which a presentation is required. Two of these have been indicated, Concept and Implementation. Blanks are included for any other decisions that must be sought. Key reviews should be included even though they will be handled entirely with the *Workbook*.

Down the left side of the matrix is a list of management levels to which presentations might have to be given. To the right of the management levels and under each column heading designating a decision needed, are spaces for entering the number assigned to the presentation and the tentative date. The numbers serve to cross-reference the matrix entries to the data at the bottom of the work sheet. If a presentation asking for approval of the concept must be given first to middle management and, at a later date, to upper management, each occurrence should get a different number. The reason for planning each occurrence of the same presentation separately is that the analyst must anticipate different sets of questions from the different audiences.

On the lower part of the work sheet, the analyst should fill in as much data as possible for each presentation, following the numbering scheme used above.

After "To Whom," the analyst should indicate the management levels that will be present. The next blank can be used for indicating the specific action desired.

After "Major Criteria," the analyst should indicate the most important factors key decision makers will use to make their decision. If reference material is needed, it should be identified in the "Back-Up Needed" portion of the work sheet.

THE PRESENTATION AUDIENCE WORK SHEET

As key decision makers are identified, the analyst should start filling out a Presentation Audience work sheet for each of them. This work sheet is shown in Figure 137. This process helps the analyst develop a good understanding of what to anticipate from his audience, enabling him to be better prepared with the proper material.

After filling in the person's name, title and organization, an indication should be made whether he or she has a direct decision-making role or an advisory role. There should also be an indication as to whether this person's responsibilities relate to technical or operations aspects of the system.

In entering the information on how this decision maker may be affected by the problem and the proposed system, the analyst should be sure to frame the description from the decision maker's point of view. A manager in charge of a function that would use the new system may see the problem as one of efficiency; the manager of a department that is affected by the system may see it in terms of timeliness of data.

The most difficult and also most important task in filling out this work sheet is anticipating the questions most likely to be asked during the presentation. There is nothing more damaging than to be caught tongue-tied by an unanticipated question. This can usually be avoided or minimized by finding out, in advance, as much as possible about each potential member of the audience, including background and interests. Anticipated

PRESENTATION AUDIENCE 8.19

Name____________________ Title/Organization ________________

Decision-making Role:

____ Direct or ____ Advisory

Responsibility:

____ Technical or ____ Operations

How Affected by Problem________________________

How Affected by System ________________________

What are Seen as Principal Benefits ________________________

Anticipated Questions	Answers	Back-Up	
		Work Sheet	Other

FIGURE 137: Presentation Audience Work Sheet

questions can then be noted, followed by notations as to logical responses the analyst can make to the questions. Reference should also be made to back-up material where appropriate.

THE PRESENTATION PLANNING WORK SHEET

When the audience has been defined, the analyst should fill out the Presentation Planning work sheet, as shown in Figure 138.

The first item to be recorded is a description of the presentation's purpose. This is followed by a notation as to the specific action to be requested.

PRESENTATION PLANNING **8.20**

Purpose (Decision Desired) ______________________ Date ____________

____ Place ____________

Specific Action to be Requested ______________ Type ____________

____ Presentation No. ______

Audience:

Primary Decision Maker ______________________________________

Other Attendees ______________________________________

____

____

Primary Interest of Audience ___ Technical; ___ Operations; ___ Costs/Profits

OUTLINE

Key	Item	Graphics	Back-Up

FIGURE 138: Presentation Planning Work Sheet

In identifying the audience, the analyst should indicate who will make the ultimate decision. Usually, but not always, this will be the highest management-level person attending. Other expected attendees should also be listed, with a separate Presentation Audience work sheet made out for each of them, as previously described.

The bottom half of the work sheet is used for making an outline of the presentation and listing, in the process, the key points that should be made. At this point, the serious job of summarizing begins by selecting which items must be included and which can be left out. The outline should be checked for logical flow and for potential transitions between one point and the next. Does the logic really culminate in a convincing reason for the action wanted? Are the items included essential to the logic? Finally, the analyst should check the outline against the questions which have been anticipated and noted on the Presentation Audience work sheet.

Columns are also provided on the Presentation Planning work sheet for identifying graphics that will help illustrate key points. There is also a column for referencing appropriate back-up data.

19
Documenting the System

The cost of operating a massive computer-based information system recently began to skyrocket for a large West Coast organization. This, plus mounting evidence that the function was being poorly administered, prompted management to launch an investigation. What they found dismayed them.

They found that the system, although in its eighth year of operation, existed *without a single shred of system or program documentation!* In the beginning, their manager in charge of this function had contracted with a consultant-friend of his to design, install and operate the system. Provisions for documentation had apparently been intentionally omitted from the agreement. From a practical standpoint, as things now stood, no one except these two could operate, evaluate, fix, change, or improve the system. The essentials of the system's design and operation were in their heads (or in clandestinely maintained notebooks). This was their approach to maintaining job and income security. Needless to say, all of the solution options available to management, in this particular case, were costly ones.

THE NEED FOR DOCUMENTATION

The need for a system to be documented is obvious. First, of course, is the need for a specification manual where the system and its programs are fully described and explained. This type of manual is essential for trouble-shooting purposes, for modifying or upgrading the existing system, and for designing a new one.

There is also a need to provide instructions to the people who will operate the system. This usually takes the form of an operations manual for the mechanized data

8.21

PROCEDURE PLANNING

Ref. ____________________

System ____________________

System Segment to be Covered by This Procedure ____________________

____

Who Needs the Procedure? ____________________

____

What Information Should it Contain? ____________________

____

____

____

____

How Much Detail is Needed? ____________________

____

____

____

____

Which Departments will be Affected by This Procedure? ____________________

____

____

In What Form will the Procedure be Issued? ____________________

____

____

Will Employee Indoctrination or Training be Necessary? ____________________

Describe ____________________

____

FIGURE 139: Procedure Planning Work Sheet

processing portion of the system. The more familiar procedures manual is usually the vehicle for conveying the instructions for the manual portion of a system's operation.

There can also be manuals or instructions on how to use a system. This is especially true of consumer-oriented systems. Other types of system documentation could include forms manuals, sales manuals and training manuals. An essential thing to remember is that all the documentation for a given system is related. A change in one could easily affect the others.

In all cases, the system documentation should be clear and easily understood by others. There should also be a method for maintaining it so that it always reflects current conditions.

OPERATING PROCEDURE (Work Sheet) 8.22

Subject: Processing of Mailed-In Price Queries
Procedure No. P-176
Page 1 of 3
Approved: R. White
Date: May 23
Department(s) Affected: Dept. 20 Mail Service Unit; Dept 35 Sales Quotation Department; Dept 50 Order Fulfillment Department.

Reference(s): Flowchart (attached)

Purpose: To provide customers, and potential customers, with personalized, friendly, accurate and quick responses to their mailed-in requests for pricing information.

SEQ.	PROCEDURE STEP (Use Continuation Sheets, as Req'd)
1.	Letters received by Department 20 Mail Room are opened and sorted by Mail Service Unit as to orders (next step: Sequence 2) and queries (next step: Sequence 3).
2.	Orders for Company products are batched twice daily and delivered to Order Fulfillment Department 50 (NOTE: This is a separately defined

(continued on next sheet)

FIGURE 140: Example of Use of Operating Procedure Work Sheet

THE PROCEDURES MANUAL

Procedures describe specific courses of action or ways of doing things. Assembled into manuals, they are the official instruction books for non-mechanized systems, or for the manual portions of mechanized systems.

An essential first step in the procedure process is their planning. A work sheet that can be used for this purpose is shown in Figure 139, Procedure Planning work sheet.

PROGRAM SUMMARY **8.23**

File No. ____________________

Program Name ____________________

Program Identification Number ____________________

Application ____________________

____

Designed By ____________________ Date ____________

Approved By ____________________ Date ____________

Language ____________________

Computer ____________________

Input Devices ____________________

____

Output Devices ____________________

____

Run Frequency ____________________

Program Description ____________________

____

____

____

____

____

____

____

____

____

____

FIGURE 141: Program Summary Work Sheet

Information for this can be derived from the Manual Operations Procedure Requirements work sheet described earlier in Chapter 11.

This particular sheet calls for the identification of the system and the system segment that is to be covered by the procedure. Questions asked on the form focus attention on an identification of who needs the procedure, what information it should contain, and how much detail is needed. The procedure planner is also asked to consider which units in the organization will be affected by the procedure, the form in which it will be issued, and the possible need for employee indoctrination or training. New ways of doing things often require special educational programs of one type or another.

A specially designed work sheet can be useful in preparing rough drafts of a procedure. One such form is shown in Figure 140, Example Use of Operating Procedure work sheet. Input for this can be derived from the New System Flow work sheet described in Chapter 11. The form example shown in Figure 140 is filled out to show how it could be used in an actual application. One interesting notation is the reference to a flowchart. Manual and mechanized processes both benefit from the graphic clarity of a well-drawn flowchart. (See Chapters 8 and 11.)

DOCUMENTATION OF AUTOMATED SYSTEMS

A summary description of a data processing program can be maintained on a work sheet such as the one shown in Figure 141, Program Summary work sheet. This is merely the tip of the iceberg though, for the normal automated system and its component programs require a great deal of documentation. Included in such documentation would be flowcharts, logic descriptions and explanations, program listing, record and report layouts, operating instructions and other items.

A comprehensive listing of these items has been assembled here and divided into two sections: those needed for a specification manual and those needed for an operating manual.

Specification Manual

A checklist is combined with a work sheet in Figure 142, System/Program Specification Manual Checklist and Assignment Sheet. It can be used as a checklist to indicate those items that should be included in a particular specification manual. If the project is a large one, with a number of individuals involved in the development work, then the same form can also be used to assign individual responsibility for the preparation of each selected item and to assign review responsibility. An alternative would be to use the right-hand column for specifying the date when a completed version is due. The following are brief descriptions of the contents of a typical specification manual, as shown on the checklist.

Cover, Title Page. These two items should contain the official title of the system, the system's identification number, the name and location of the system designer, the date when the system and programs became operational, and the appropriate approval signatures.

Table of Contents. This should contain a complete listing of the manual's contents with page number references.

Change Control Log. This would be a sheet for maintaining a record of changes. Spaces should be provided for referencing the control numbers of the change notices (copies of these notices could also be attached), for recording the dates when the changes were made, for identifying the programs affected by each change, and for identifying the names of the persons who made the changes.

System Purpose. This should be a "layman's language" description of the purpose and objectives of the system.

Processing Description. This is a narrative description of the system's processing steps. This would include a description of the input data, the ways in which the data would be manipulated, a description of the various files, and the types of output produced. This description should be related to the program processing description which would follow later in the manual.

System Flowchart. The specification manual should contain a comprehensive system

8.24

SPECIFICATION MANUAL

Checklist and Assignment Sheet

System Title ______________________ No. __________

Item	Req'd		Assign	Review
	Yes	No		
Cover				
Title Page				
Table of Contents				
Change Control Log				
System Purpose				
Processing Description				
System Flowchart				
Run Sequence/Options				
File List				
Data Element Cross-Reference				
Description of Reports				
Report Layouts				
Record Layouts				
Program Purpose				
Program Processing Description				
Limitations and Restrictions				
Program Flowchart				
Decision Tables				
Program Listings				
Glossary				
References				
Exhibits				

FIGURE 142: System/Program Specification Manual Checklist and Assignment Sheet

flowchart which graphically displays both manual and machine processes. (See Chapter 11.)

Run Sequence/Options. Many systems contain the capability of producing various optional outputs, such as error-correction runs, regularly scheduled runs and special request runs. This sheet identifies, for each option, the appropriate combinations and run sequences of the system's various programs.

File List. Called for here is a sheet for identifying a system's files in terms of title, medium (such as cards, tape, disk pack), and reference number.

Data Element Cross-Reference. Data elements that are used in the system are cross-referenced, on this sheet, to the specific files where they are used. Each element is identified as to title and reference number, description, the number of card or record positions required, an identification code as to whether it consists of alpha, numeric, or special characters, or a combination of these, a coded indication as to whether it is right or left justified, and a reference to the file, or files, where it is used.

Description of Reports. A separate sheet for describing each report used in the system and generated by the system is called for in this section of a specification manual. Space should be provided on each sheet for identifying the report in terms of title, number, description, form number, frequency of generation and number of copies. A distribution list, with addresses, can also be given. In the case of computer-generated output reports, instructions as to whether or not the output should be burst and/or decollated can also be recorded.

Report Layouts. Examples of reports used in the system, together with layouts of computer-generated reports showing the spacing of report fields, are displayed in this section of the specification manual. Each report should be identified as to title, report number and data set source.

Record Layouts. This section of the manual is used for identifying the system records, referencing them to the appropriate files and describing and displaying their layouts.

Program Purpose. The purpose of each program used in the system should be described here. This would include an explanation of each program's relationship to the rest of the system.

Program Processing Description. This is a detailed description of the processing steps required in each program, including a discussion of the processing logic and program organization, and a comprehensive explanation of any special techniques that have been used. The program processing description should be related to the system processing description presented earlier in the manual.

Limitations and Restrictions. A precise description of limitations and restrictions regarding the system's applications should be presented in this section of the specification manual.

Program Flowchart. This section contains the graphic portrayal of each program used in the system, showing the logical sequence of the processing steps.

Decision Tables. Decision tables used to graphically display processing option conditions are displayed in this section of the specification manual.

Program Listings. This is a listing of the program statements, showing the coded instructions and their sequence.

Glossary, References, Exhibits. These sections of a specification manual are used for explaining and defining technical terms or abbreviations, for listing and citing special references that are germane to the system, and for displaying appropriate exhibits.

Operating Instructions Manual

A combination checklist/assignment sheet for an operating instructions manual is shown in Figure 143. The following are brief descriptions of the items listed on the checklist.

8.25

OPERATING INSTRUCTIONS MANUAL
Checklist and Assignment Sheet

System Title ______________________ No. ________

Item	Req'd		Assign	Review
	Yes	No		
Cover				
Title Page				
Table of Contents				
Change Control Log				
Schedules & Time Data				
Input Instructions				
Balancing Instructions				
Keyboard Instructions				
Tabulating Flowcharts				
Control Cards and Tables				
Run Sheets				
Program Flowchart				
Messages				
Program Reentry/Restart				
Output Exhibits				

FIGURE 143: Operating Instructions Manual Checklist and Assignment Sheet

Cover, Title Page. As with the previously described specification manual cover and title page requirements, these items should contain the official title of the system, the system's identification number, the name and location of the system designer, the date when the system and programs became operational, and the appropriate approval signatures.

Table of Contents. This should contain a complete listing of the manual's contents together with page number references.

Change Control Log. This is a log sheet for maintaining a record of changes, including the dates changes were made and the names of the individuals who made the changes. Change notices can be referenced by number with copies of the notices appended to the log sheet.

Schedule and Time Data. Run frequencies and dates, turnaround times and estimated processing times for the standard production runs and for the various program options are recorded in this section of the operating manual.

Input Instructions. Examples of transmittal and input control forms are displayed in this section, together with instructions to the data processing center as to the control operations to be performed on the input.

Balancing Instructions. Instructions are given in this section for balancing the system and for corrective action when balance is not attained.

Keyboard Instructions. Keyboard instructions, as well as examples of the source documents, are displayed in this section of an operating manual.

Tabulating Flowcharts. When tabulating machines are used in the system, flowcharts which graphically display the steps involved can be shown in this section.

Control Cards and Tables. This section of the operating manual can be used for providing instructions and formats of control cards and tables, and for providing the operator with instructions as to the arrangement of cards within a deck.

Run Sheets. Production and check-out run sheets can be shown in this section. Information on the sheets includes instructions to the operator regarding loading and running the program.

Program Flowcharts. Charts which graphically show the flow of data through each of the programs can be displayed in this section.

Messages. This section is for listing the program-generated messages that the operator is likely to encounter when running the program, and for citing the causes and responses that should be made.

Program Reentry/Restart. This section is for use in providing program reentry routines and restart instructions.

Output Exhibits. This section is used for displaying examples of system output.

20
Monitoring and Evaluating System Operations

The system is "up and operating," but is it accomplishing its intended purpose? That is the central question asked after a system has been designed and implemented. It is a question that can be answered only if the new system's operations are being monitored and evaluated. If an evaluation reveals that the system is not performing properly, the problem, or problems, should be identified and the system modified or redesigned accordingly.

THE MANAGEMENT OF CYCLIC OPERATIONS

Up to this point in the book, the management techniques described have been those necessary for planning and controlling a one-time activity. Such an activity is one that consists of sets of tasks that are performed only once, such as those required to build a bridge, construct a building, design a prototype, or design and implement an improved system of operations.

Most other types of activities, whether operating shoe stores or manufacturing companies, are cyclic in nature. In other words, there is a certain repetitive pattern to the work that is being done. These types of activities require a set of management tools different from those required for a one-time project. Cyclic operations need:

- A description and quantifying of goals (quotas, schedules, quality and performance parameters).

SYSTEM EVALUATION FACTORS 8.26

Checklist

Factor	Check (✓)	
	Routine Tracking	Random Analysis
PRODUCTIVITY		
Schedule Performance		
Queue Time		
Cycle Time		
Workload vs. Capacity		
Other		
COST		
Budget Performance		
Unit Cost Analysis		
Other		
SYSTEM OPERATION		
Maintenance Record		
Trouble-Shooting Record		
Documentation Audit		
Other		
ERRORS		
Frequency		
Magnitude		
Other		
EMPLOYEE MORALE		
Grievances		
Absenteeism		
Tardiness		
Turnover		
Other		
USER REACTION		
Satisfaction		
Complaints Regarding Service		
Complaints Regarding Qual./Accuracy/Rel.		
Other		

FIGURE 144: System Evaluation Factors Checklist

- A description of the operating steps required to achieve these goals (operating procedures, instructions, flowcharts).
- An allocation and budgeting of resources (manpower, equipment, facilities and materials) required to achieve these goals.
- An assignment of authority and responsibility for managing the overall operations, including tracking actual performance in relationship to planned performance.

TIME EVALUATION 8.27

System ____________________ Date __________

Work Element	Queue Time			Cycle Time		
	Old	Plan.	Act.	Old	Plan.	Act.

FIGURE 145: Time Evaluation Work Sheet

SYSTEM EVALUATION FACTORS CHECKLIST

The specific factors that should be monitored and evaluated will vary greatly depending on the type of system. A further consideration is that once a selection of factors has been made, it must be decided which are to be tracked on a regular, routine basis, and which should be made the subject of random analysis only.

The System Evaluation Factors Checklist shown in Figure 144 can be used as a

8.28

COST EVALUATION

PER (CIRCLE ONE): HOUR DAY WEEK MONTH

System ______________________________ Date ______________

Cost Factor	Old System	New System		
		Planned	Actual	Variance
Manpower[1]				
Materials[2]				
Equipment[3]				
Facilities[4]				
Other:				
Other:				
Other:				
Other:				
Totals				
Jobs or Units Processed				
Cost Per Job or Unit				

Types, Quantities and Rates

[1]Manpower	
[2]Materials	
[3]Equipment	
[4]Facilities	
Other:	
Other:	
Other:	
Other:	

FIGURE 146: Cost Evaluation Work Sheet

starting point in making these determinations. A basic list of factors that might apply to many systems is shown. Extra spaces have been provided as well, so the analyst can add additional factors appropriate to a specific system.

Time, schedule and workload aspects of a system are listed first. The monitoring of schedule performance involves comparing actual versus planned accomplishments performed during specific time periods. Ideally, the measurements are made on the basis of the individual steps involved in the process.

A monitoring of system queue time involves tracking the length of time a job must wait before being processed and evaluating whether that time is excessive or not. Cycle time can be a measurement of the process time only, or it can be defined as the total "turnaround" time. Workload can be continually monitored against the system's rated capacity as one approach to trying to anticipate problems before they occur.

Two types of cost monitoring are listed on the checklist. The monitoring of budget performance involves tracking actual expenditures versus planned expenditures. An analysis can also be made to determine the unit cost of the product or service on either a continual or intermittent basis.

There are a number of ways that system operations can be monitored and evaluated. As one approach, a maintenance record can be kept along with a record of troubles encountered during system operations. Another thing that can be done is to periodically audit the system's documentation to see if it is up to date, accurate and adequate. Errors can be tracked, too, both in terms of frequency and magnitude.

Employee morale can be an important aspect of certain types of systems. In those cases where it is, thought should be given to monitoring and evaluating this condition from the standpoint of grievances, absenteeism, tardiness, employee turnover or other factors.

Of equal importance to the operation of certain types of systems are the reactions of users. In these cases consideration should be given to monitoring and evaluating user satisfaction, complaints regarding service, quality, accuracy, reliability and other similar factors.

EVALUATION WORK SHEETS

There are numerous formats and chart styles that can be used to monitor and evaluate each of the many elements of a system's operation. The analyst must choose from this huge array of options those that are best suited to the system he or she has designed and implemented.

As a starting point, two work sheets are displayed here. The first, Time Evaluation work sheet, Figure 145, provides, on the left, a column for identifying each of the separate steps involved in the system's process. Columns are then provided for recording queue times and cycle times. Actuals for each category can be compared to what has been planned, as well as to the times that were required for the same steps in the old system.

Figure 146, Cost Evaluation work sheet, provides a format for analyzing actual versus planned costs for each of a number of separate cost factors. As with the previously described work sheet, columns are provided for comparing current costs with those of the old system. The bottom portion of this particular work sheet provides spaces for further explaining the cost factor items that have been listed.

21

Recapitulation: Tying the Workbook System Together

Systems analysis is a craft which usually requires equal amounts of inquisitiveness and creativity, on the one hand, and patience and attention to details on the other. Although not all system improvements are "innovations," the systems analyst's ingenuity is usually challenged, nonetheless. But before imagination can be applied, the existing system and the environment in which it operates must be thoroughly understood; before the appropriate change can be made, all the available alternatives must be thoroughly examined. This requires an exhaustive, and often tedious, compilation of facts.

This book has presented a systematic approach to the collection and recording of these facts. The work sheets and checklists cover all of the areas which an analyst must investigate for most systems projects. When assembled, the Workbook, itself, becomes a master checklist for the project. There is greater likelihood that the Workbook will cover more items of information than are needed instead of too few. Thus the analyst, in selecting the items that are significant, will be most likely omitting items through conscious decision, rather than through neglect.

As he completes each work sheet, the analyst will have confidence that his task has been properly completed. If there are gaps in the data, it will be because that information is not available or is not pertinent to the task at hand. Furthermore, he will have automatically organized data for its use in subsequent analysis tasks.

For jobs of lesser scope or of a more specialized nature, the analyst, after reviewing the whole Workbook, may select only those work sheets and checklists that are pertinent, and he may add and delete items or prepare supplementary work sheets of his own. By reviewing, modifying, and assembling the Workbook in advance, he will have planned his data gathering and analysis beforehand.

The work sheets and checklists presented here are, therefore, useful tools in the conduct of a systems analysis whether by a team or by an individual, whether of a small-scale manual system or of a large and complex automated system.

The Workbook has another value when used by a team of analysts of diverse backgrounds. With all the team using work sheets of the same format, consistency is assured. In addition, as the project proceeds, each analyst will be able to find the information appropriate to his task regardless of who has recorded it. He will have reasonable assurance that all of the information he needs will have been recorded, regardless of the specialty of the recorder.

In assembling a Systems Analysis Workbook, it should be arranged for maximum usefulness during the analysis and new-system development phases. This may mean some skipping around when recording data, but this is much to be preferred to frustrating and time consuming searches later. As presented in this book, however, some selections were discussed in a different sequence to provide a more coherent explanation. Therefore, a review of the Workbook contents as it would normally be assembled is presented below in this chapter. In this listing, the individual work sheets are cross-referenced to the figures in the text where they are explained. The references are to figure numbers; the explanatory materials will be found on adjacent pages. As explained in Chapter 2, the basic organization of the workbook is indicated by the nine tabs, each designating a major category of data. The numbering of individual work sheets is based on the tab numbers, the first digit of the sheet number being the number of the tab. This is followed by a decimal, and the digit that follows is the sequential number of that work sheet.

The asterisks in the list indicate the work sheets used in one application of the Workbook to a small-scale project for the improvement of a manual system. It was a system for tracing rejected materials, a one-man study effort requiring about three weeks. The system was entirely manual and was centered about the processing of a single form. The objective was to reduce the number of copies and the handling of this form. No sheets from Tab 8.0, Implementation Plan, were used since implementation consisted simply of writing one procedure, obtaining approval, and having it issued.

TAB 1.0 Analysis/Design Plan

Although not always thought of as part of the analysis, planning is essential to the success of a systems project, and in fact does constitute the beginning of the analytical process. At the very least, the objectives, constraints, and scope of even a small project must be precisely understood and documented in the two specification work sheets. The remaining work sheets provide for more detailed planning data. (See Chapters 2 and 4.)

Sheet 1.1A	Workbook Contents Checklist (Sheet 1 of 2)	Figure 6A
1.1B	Workbook Contents Checklist (Sheet 2 of 2)	Figure 6B
1.2*	Project Specification—Scope and Limitations	Figure 7
1.3*	Project Specification—Schedule & Location Considerations	Figure 8
1.4*	System Specification	Figure 9
1.5	Objective Considerations Checklist	Figure 10

Sheet 1.6	Task Planning—Analysis & Design Plan	Figure 12
1.7	Resource Estimating—Analysis & Design Plan	Figure 17
1.8	Task Assignment—Analysis & Design Plan	Figure 20
1.9	Design Review Checklist	Figure 19
1.10	Interview Planning Checklist	Figure 30

TAB 2.0 System Environment Factors

This section of the Workbook is used for recording information about the environment in which the system must operate. This is not data about the system, *per se,* but it is data which is essential to the understanding of the system, of why it operates the way it does. The environment often determines the limits or constraints on how the system can be changed. (See Chapter 3.)

Sheet 2.1*	Organization Identification	Figure 21
2.2*	Sub-unit Identification	Figure 22
2.3*	Key Personnel	Figure 23
2.4	Headcount	Figure 24
2.5	Plans and Trends	Figure 25
2.6	Historical Data	Figure 26
2.7	Equipment Description—Existing Data Processing Hardware	Figure 27
2.8	Equipment Description—Data Communications, Reproduction, Business Machines	Figure 28
2.9	Equipment Description—Data Processing Hardware Order	Figure 29

TAB 3.0 Existing-System Flow

Under this tab is collected information on how the existing system works, including data on the manual procedures used, the computer programs that exist, and those programs under development. It also includes data on which the existing-system flowchart is based. (See Chapters 6 and 8.)

Sheet 3.1*	System Flow	Figure 40
3.2*	System Features Checklist	Figure 42
3.3	Procedure Summary	Figure 43
3.4	Automated System Identification	Figure 44
3.5	Automated System Breakdown	Figure 45
3.6	Automated System Summary	Figure 46
3.7	Developmental Program Status	Figure 47

TAB 4.0 Existing-System Documents

The information processed by a system is represented by the documents used for input, the documents produced, the system files (manual or automatic), and the data elements contained in these files and documents. This tab provides work sheets for recording data on the contents and format of this information. Actual examples of each document should be collected and filed under this tab as well. (See Chapter 7.)

Sheet 4.1*	Existing-System Document Identification	Figure 48
4.2*	Existing-System Manual File Description	Figure 49
4.3	Existing-System Data Base Description—Automated Files	Figure 50
4.4	Data Element Inventory	Figure 51
4.5	Data Element Matrix	Figure 52

TAB 5.0 Requirements

The requirements a system must meet are established primarily by its objectives. These functional requirements are shaped and added to by the environment, particularly by the organizational relationships, and constrained by overall policies and regulations. Finally, the people who use the system will impose requirements and have preferences which must be recognized if the system is to be successfully operated and used. (See Chapter 5.)

TAB 6.0 Evaluation Criteria

Before a change to a system can be justified, the benefit to be derived must be established as well as the cost of making the change and the increased operating costs, if any. Therefore, the existing operating costs and characteristics of the existing system must be established to provide a baseline for comparison with costs and benefits of the proposed new system. This section contains the work sheets for recording needed raw data and for translating this data, as it relates to both new and existing systems, into comparable quantities. (See Chapter 9 for existing system analysis; Chapter 15 for proposed system evaluation.)

TAB 7.0 New-System Design

The title of this tab is self-explanatory. The new system concept is described in sufficient detail to define its significant details and to permit planning and carrying out the next step, detailed design and implementation. (See Chapters 11, 12, 13 and 14.)

TAB 8.0 Implementation/Documentation/Evaluation

A well-conducted systems analysis is normally followed by detailed design and implementation of the new system. This means programs must be developed and tested for automated systems (if the new system requires a computer), procedures must be written, equipment acquired and installed, and the operators and users trained. This tab contains the work sheets for recording the detailed planning data necessary to implement the new system efficiently, to document it, and to evaluate results. (See Chapters 16, 17, 18, 19 and 20.)

Sheet 8.19	Presentation Audience	Figure 137
8.20	Presentation Planning	Figure 138
8.21	Procedure Planning	Figure 139
8.22	Operating Procedure	Figure 140
8.23	Program Summary	Figure 141
8.24	System/Program Specification Manual Checklist	Figure 142
8.25	Operating Instructions Manual Checklist	Figure 143
8.26	System Evaluation Factors Checklist	Figure 144
8.27	Time Evluation	Figure 145
8.28	Cost Evaluation	Figure 146

TAB 9.0 Glossary/Appendix

This section is for compiling a list of terms or abbreviations unique to the system under study, and for accumulating other types of useful reference materials related to the project. (See Chapter 2.)

Sheet 9.1	Glossary	Figure 11

Index

A

B

C

T

U

V

W